The Formative Years of the Telegraph Union

This book is a collective effort of the research group involved in the Swiss National Fund Project entitled "The Swiss Influence in the ITU's decision-making processes 1865-1914". Sources and ideas were discussed collectively, while the chapters were written individually. Giuseppe Richeri wrote the Introduction, Gabriele Balbi Chapter 3, Simone Fari Chapters 1,2,4,5 and the Conclusion.

The Formative Years of the Telegraph Union

By

Simone Fari
With contributions by

Gabriele Balbi and Giuseppe Richeri

Translated by Patricia Kennan

Cambridge
Scholars
Publishing

The Formative Years of the Telegraph Union

By Simone Fari
With contributions by Gabriele Balbi and Giuseppe Richeri

This book first published 2015

Cambridge Scholars Publishing

Lady Stephenson Library, Newcastle upon Tyne, NE6 2PA, UK

British Library Cataloguing in Publication Data
A catalogue record for this book is available from the British Library

The translation of the book has been realized by Fondazione Hilda e Felice Vitali (Lugano)

ISBN (10): 1-4438-7706-9
ISBN (13): 978-1-4438-7706-0

CONTENTS

ACKNOWLEDGEMENTS

In 2009 a closely-knit research group consisting of Simone Fari, Gabriele Balbi and Giuseppe Richeri, all working at the Institute of Media and Journalism of the Università della Svizzera italiana in Lugano, set up a long-term project entitled "The Swiss Influence on the ITU's decision-making process 1865-1914". As the title indicates, the aim was to analyze the role played by Switzerland in the creation, early years and decision-making processes of the Telegraph Union (known today as the International Telecommunications Union), and bring together a host of political, economic, technical and cultural points of view. As our work progressed, new and unexpected factors started to emerge, bringing us to formulate a re-interpretation of the formative years of the TU, and the result is quite simply this text.

Over the years in which we have been involved in such a specific and complex piece of research, we have collected many debts, both of an intellectual and material level. We would like to thank in particular Verdiana Grossi, Andrea Giuntini, Jakob Tanner, David Gugerli and Richard John, as well as all the participants on the panel at the ICOHTEC Annual Conference in Barcelona 2012 (especially Simone Müller, John Laprise and Colin Hempstead). We would also like to mention the panellists of our session both at the HISTELCON Conference in Pavia 2012 (in particular Anna Guagnini and Jonathan Coopersmith), and at "Tensions of Europe", Paris 2013.

We have debts with Heather Heywood and all the personnel of ITU Library and Archives, and we would like to thank the USI-SUPSI Research Service (Lugano), for their support in the pre-writing phase of the project and for their help in submitting it. Our greatest debt is to the Swiss National Fund, which financed our project right from the beginning, for without its help and backing this research would certainly never have seen the light.

INTRODUCTION

In the second half of the nineteenth century, the powerful development of an international telegraph network laid the infrastructural base for what can be termed a first moment of globalization. The dynamics of this network's growth were very much determined by Great Britain, the major politico-economic powerhouse of the period, and the balance of interests it struck up with up-and-coming economies, especially the US and Germany. The control room of the network was in the hands of the first ever international organization, the International Telegraph Union (TU), which was responsible for the supranational regulation of communications. Its official task was to execute the dictates of the single national governments and carry out purely administrative duties. But is that what actually happened?

Though research has been carried out on the subject, there has been up to now no reconstruction of the TU's origins. Historical analyses most often concentrate on the period post 1865, the year in which the TU was instituted, and leave aside the chain effect the initial choices of the previous twenty years had on the organization's structure and decision-making processes. Very often, too, such studies concentrate on secondary sources, and tend to pass over primary ones like the documents of the TU and its predecessors.

One of the characteristics of the present study is to reconstruct meticulously the events that led to the telegraphic multilateralism of the period 1849 to 1875. The dates are significant – in 1849 the first telegraph treaty was signed, so giving birth to telegraphic multilateralism, while 1875 is the date of the last TU diplomatic conference, which was responsible for the Convention of St Petersburg, the constituent, inalterable act of the TU until post WWI.

The approach used in the present study is institutionalist, and privileges sources generated by the Union itself (conference minutes, correspondence of the permanent body, articles from official revues and other administrative documents), without neglecting intertextual analysis (parliamentary bills/acts from various countries, articles in international revues, technical and grey literature). The numerous interpretations of the shadow areas in the history of the TU we propose have therefore a solid base in sources, which is the reason why this study is backed by such an abundance of detailed footnotes.

The methodology used in reconstructing historical events is multidisciplinary, representative of the members of our group, who come from various scientific approaches and possess different skills and competences. This text, the result of over three years' research, is therefore simultaneously a history of media studies, international relations and business.

The aim of this work is first of all to answer some questions which have not until now received satisfying answers. For example, who was really at the head of the TU? Was it the governments or the elite state officials? What role did private companies play? How far did the TU determine the standardization of norms and technology along international lines? How were international tariffs established? What was the logic underlying them? What vision of an international service emerges from the telegraph conferences?

The first chapter explains how and why the TU can be considered the final result of the development of telegraph multilateralism in Europe, which started with the first bilateral treaties in 1849 and then evolved into the two unions, one Francophile, the other German-speaking.

The second chapter starts with a description of the bodies created by the TU and how the way in which they operated influenced the modes of action of the Union itself. It investigates both the players and the processes of elaborating and approving norms which characterized the telegraph conferences until 1875. The third chapter goes on to consolidate our argument by illustrating the tasks and real power of the Bureau, the TU's permanent body, which was originally planned to carry out a purely administrative role, though it ended by overcoming the limits which had originally been set.

Chapters Four and Five are both dedicated to the three main functions of the TU - the standardization of general norms, technology and tariff uniformity. Chapter Four in particular discusses how the process of standardizing general norms, very much influenced by the liberal ideas of the period, took up more time, space and energy than technological standardization, which, all things considered, was always of secondary interest. Chapter Five, instead, untangles the confusion of interpretations, schools of thought and business strategies which came to bear on the application of uniform tariffs. It also describes how the various ideological currents underlying different criteria of charges corresponded to different concepts of the telegraph service.

Finally, after having dissected the TU's structures and detailed its basic elements in the various chapters, we come to the conclusion, which furnishes an overall vision of our study and highlights how many of the long-term features of the TU were the consequences of a series of initial choices which in a thirty-year time span ended by stifling any attempts to be right on the cutting edge.

CHAPTER ONE

THE LONG GESTATION

Paris, 1865. After a series of meetings lasting over two months, the representatives of 20 European countries[1] came to sign the Telegraph Convention, the treatise which instituted the International Telegraph Union (TU)[2]. It was the first supranational government organization to come into existence and at the same time the first organization to oversee and regulate an international public service. It marked the arrival point of a long diplomatic journey which had begun back in 1848 and worked its way forward by means of a myriad of bilateral and multilateral treatises. The Paris Convention was, in fact, a compendium of all the norms of the international telegraph service which had been agreed upon by the various European countries over twenty or so years. Consequently, the question "Why was the TU established?" exacts the obvious answer "To sort out the confusion of norms and regulations that had piled up and overlapped over the years". But this answer begs another question: "Why did the telegraph service set off an intense diplomatic process which step by step led to the establishment of the first international organization in history?"

The answers are many, and they all lead back to an in-depth analysis of the events taking place between 1849 and 1865, years which can be considered to all effects and purposes the period of gestation of the TU. There emerged in those years some constitutive choices[3] destined to

---

[1] In alphabetical order: 1) Austria, 2) Baden, 3) Bavaria, 4) Belgium, 5) Denmark, 6) France, 7) Greece, 8) Hamburg, 9) Hanover, 10) Holland, 11) Italy (formerly Piedmont/Sardinia), 12) Portugal, 13) Prussia, 14) Russia, 15) Saxony, 16) Spain, 17) Sweden-Norway, 18) Switzerland, 19) Turkey, 20) Württemberg. The most important absentees were the UK and the US, countries in which the telegraph service was in the hands of private companies. The requisite for joining the TU was in fact a publicly managed telegraph service.

[2] *Documents diplomatiques de la Conférence télégraphique internationale de Paris* (Paris : Imprimerie Impériale, 1865).

[3] In this case the term *constitutive choices* takes on a precise methodological value. Following on from Paul Starr's analysis, it is assumed that especially in the field of the media, initial political and economic choices influence the long-term trajectory

characterize the TU at least up to well after WWI: 1) an intrinsically international character; 2) a strong political interest in the international management of the service; 3) a marked Eurocentrism (meaningful the absence of the US), with a continental hegemony foretelling a kind of premature European integration; 4) a strict adhesion to the economic and political values of liberalism, detectable not only in the conventions and their rules but also in the minutes of the assemblies discussing them.

1.1 The diplomatic imbroglio

With due differences, the European telegraph service became available to the public between the late 1840s and the mid-1850s[4]. It may seem surprising, however, that the first international treatises go back to 1848-49, and in many cases actually anticipate the setting up of a national service[5].

The real pioneer was Prussia. Many of the regions in its territory were scattered, and could only be reached by crossing other nations. This was why once its national telegraph had been planned, Prussia moved quickly to sign treatises with its various German neighbours in order to bring about domestic communications. Although the states in question were still *de facto* independent, the process of unification was well under way[6], and the telegraph agreements can be read as part of a nation-building programme destined to construct a German telegraph network as an effective start to telegraphic multilateralism[7]. For this reason all TU history scholars identify the Treaty of Vienna signed between Prussia and Austria on 3 October 1849 as the first international agreement in

---

of a technological system. Paul Starr, *The Creation of the Media. Political Origins of Modern Communications* (New York: Basic Books, 2004).

[4] Anton A. Huurdeman, *The Worldwide History of Telecommunications* (London: J. Wiley, 2003), 48-85.

[5] On the speed of the development of international telegraphy see Keith Clark, *International Communications. The American Attitudes* (New York : Columbia University Press, 1931), 90-91.

[6] For an overview of the unification of the German infrastructures see Dieter Ziegler, "Particularisme et développement des réseaux de transport dans les États allemands (1815-1866)", in *Les réseaux européens transnationaux XIXe-XXe siècles quels enjeux?*, eds. Michèle Merger, Albert Carreras and Andrea Giuntini (Nantes : Ouest Éditions, 1995), 23-32.

[7] Victor Meyer, *L'Union International des Télécommunications et son bureau* (Berne : typescript in ITU's library, 1946), 2.

telegraphy[8] and in fact it opened the way for the hard copy versions of telegrams to cross borders, so allowing Prussian and Austro-Hungarian citizens to have direct communication[9].

From then until the mid-1850s Austria, Prussia, Saxony and Bavaria all signed bilateral agreements in order to increase access to communication. These overlapping bilateral treaties ignoring network relations were addressed with the creation of the Austro-German Telegraph Union (AGTU) on 25 July 1850[10]. Initially formed by this nucleus of nations, it was first expanded to include the other German states linked telegraphically and politically to Prussia and then all states which came more generically into Austria's area of political influence (the Grand Duchy of Tuscany, the duchies of Modena/Reggio Emilia and Parma/Piacenza as well as the Papal States)[11].

Between 1850 and 1854, when the international telegraph network was growing by leaps and bounds, all the main European states felt the impelling need to join, as witness all the countless bilateral treatises recorded between neighbouring states, including: Belgium-Prussia 1850, Belgium-France 1851[12], France-Switzerland[13], France-Baden[14], and Austria-Switzerland 1852[15], Piedmont-Switzerland, Austria-Piedmont[16], Piedmont-France[17], Baden-Switzerland, France-Bavaria[18] and Switzerland-Württemberg 1853, France-Spain[19] and Switzerland-Spain 1854[20]. Nevertheless, as emerged in

---

[8] George Arthur Codding Jr., *The International Telecommunication Union. An experiment in international cooperation* (Leiden : E.J. Brill, 1952), 13.

[9] Géo Fr. Martens, *Recueil général des traitées* (Guttingue, 1856), Tome XIV, 591-595.

[10] Léopold Neumann, *Recueil des Traités et Conventions conclus par l'Autriche avec les puissances étrangères, depuis 1763 jusqu'à nos jours* (Leipzig : Brockhaus, 1859) vol. V, 196-243.

[11] The only exception Holland, which joined AGTU because of close commercial and economic ties with the Austro-Germanic area. *L'Union Télégraphique Internationale (1865-1915)* (Berne : Bureau International de l'Union Télégraphique, 1915), 3.

[12] A.J.H. De Clercq, *Recueil des Traités de la France* (Paris, 1880), Tome VI, 100-101.

[13] Ibid., 256-262.

[14] De Clercq, *Recueil des Traités*, Vol VI, 213-217.

[15] Neumann, *Recueil des Traités et Conventions*, Vol V, 592-594.

[16] Ibid., Vol VI, 107-109.

[17] Ibid., 322-324.

[18] De Clercq, *Recueil des Traités*, Vol VI, 364-366.

[19] Ibid., 473-481.

[20] Codding Jr., *The International Telecommunication Union*, 14-16.

the case of the German-speaking network, bilateral agreements were not enough by themselves to keep apace with the rapid expansion, which called for multilateral regulations and norms simultaneously recognised by a number of states[21]. And in fact already in Paris in 1852, France, Belgium and Prussia as representative of AGTU had met to sign a telegraph convention.

When the delegates from Belgium, France and Prussia, again representing AGTU, met in Berlin in order to renew the Paris Convention, the implicit understanding was to create a common area for telegraph exchange. The basis of the project was to be the approval of lower telegraph tariffs, applicable over all the territory belonging to the signees[22]. Prussia, however, did not approve of the change and gave two reasons: 1) the decision was not a part of its AGTU mandate, which had been simply to renew the 1852 Convention; 2) its network was not sufficiently developed to take on board such a tariff reduction[23]. While its first reason can be justified at least from a strictly diplomatic point of view, the second arouses some perplexity. The state-backed Prussian telegraph network was actually one of the most advanced both technologically and organizationally, as is shown by the precocity of the telegraph agreements it signed[24]. It was highly unlikely, therefore, that a reduction of international tariffs could stimulate use to the point of risking overloading the lines. Whatever, it was Prussia's refusal which pushed five non-AGTU members (Belgium, France, Piedmont, Spain and Switzerland)

---

[21] Scholars of international relations have always debated the precise definition of multilateralism. In this study it is used in its widest and most usual sense, as an agreement between three or more states. For an interesting discussion see John Gerard Ruggie (ed.), *Multilateralism Matters: Theory and Praxis of an Institutional Form* (New York: Columbia University Press, 1993).

[22] *Légation de France en Suisse à le président de la Confédération Suisse docteur Furrer*, 18 octobre 1855, in Archive Fédérale Berne, (afterwards AF), Fond E 52, Archive-Nr. 440, Band nr. 2.

[23] «Les Délégués du Gouvernement de sa Majesté l'Empereur ont essayé à Berlin de faire adopter une série de mesures destinées à réformer plus profondément le système et à le mettre plus en rapport avec les principes de bonne administration et d'équitable répartition des frais. Mais les propositions des Commissaires français auxquels le Représentant de la Belgique s'était immédiatement rallié en principe, rencontrèrent une résistance insurmontable dans le mandat restreint que le Commissaire prussien avait reçu de l'Union austro-allemande, et dans l'organisation encore trop incomplète du service télégraphique en Allemagne», Feuille Fédérale (afterwards FF), vol. 1, no. 4, 9 janvier 1856, p. 111.

[24] Huurdeman, *The Worldwide History*, 74-82.

to meet in Paris in December 1855 and create the Western Europe Telegraph Union (WETU)[25].

Historians have often presented AGTU and WETU as two contrasting organizations, the former promoted by German-speaking states and the latter by Francophiles[26]. Even those recognizing ties between countries in the two unions before 1855 have always spoken of East-West transversal relations, as though identifying two politically separate zones[27]. In reality, the chronology of the bilateral and multilateral agreements and the almost identical AGTU and WETU conventions (tariffs excepting) point to there being no real ideological contrasts between the two unions and that what WETU did was to simply bring together the group of the countries which had been left out in the cold by AGTU[28].

The two unions were indeed far closer in spirit than is held by much literature, as is also witnessed by the constant iteration of identical norms in the treaties they turned out. The 1855 Paris Convention, for example, was a clone of the June Berlin Treaty of the same year between Belgium, France and Prussia, which was in turn faithfully based on AGTU's most recent version. Yet in spite of the substantial homogeneity in regulations,

---

[25] «Une *sorte d'union télégraphiques* entre la Belgique, l'Espagne, la France, la Sardaigne et la Suisse, union télégraphique à laquelle, nous le savons déjà, se ralliera bientôt la Toscane par un acte d'adhésion, et dans laquelle seront admis successivement tous les États de l'Europe qui voudront en faire partie». Paris International Conference sitting, 5 décembre 1855, in AF, Fond E 52, Archive-Nr. 440, Band nr. 2.

[26] This contrasts emerges clearly in Patrice Carré, "Archéologie d'une Europe des télécommunications", *France Telecom* 70 (octobre 1989): 73-84. Less underscored, the same concept also appears in George A. Codding, Anthony Rutkowski, *The International Telecommunication Union in a Changing World* (Deedham: Artech House, 1982); Alfredo Descalzi, "Creación y desarrollo de la Unión Internacional de Telecomunicaciones", in *Las comunicaciones entre Europa y America: 1500-1993, Actas del I Congreso Internacional de Comunicaciones, 30 noviembre-3 diciembre de 1993* (Madrid: Ministerio de Obras Públicas, Transportes y Medio Ambiente, Secretería General de Comunicaciones, 1993), 388-389; Meyer, *L'Union International des Télécommunication,* 2-3.

[27] Codding, *The International Telecommunication Union,* 13-20; Francis Lyall, *International Communications. The International Telecommunication Union and the Universal Postal Union* (Farnham: Ashgate, 2011), 18-22.

[28] The Convention of 25 July 1850 which established AGTU provided in art. 41 for the adhesion of Germanic countries alone. «Chacun des autres gouvernements allemands non parties à l'Union peut, s'il installe des lignes télégraphiques, adhérer à l'Union télégraphique austro-allemande», art. 41 of *Traité d'État conclu le 25 juillet 1850 entre l'Autriche, la Prusse, la Bavière et la Saxe, portant création de l'Union télégraphique austro allemande.*

it has to be said that the myriad of bilateral and multilateral agreements formed what has appropriately been called a "diplomatic imbroglio"[29]. This because the only way to regulate telegraph communications in Europe was by simultaneously applying bilateral treaties with neighbours and mixed multilateral agreements (e.g. those between France, Belgium and Prussia) as well as the convention the particular union the state was member of. The situation could become critical, especially for small countries like Belgium and Switzerland surrounded by so many neighbours. It was no coincidence, indeed, that right from the early fifties both of them had made energetic diplomatic attempts to bring the two unions closer and thus realise a common telegraphic space on a formal level. Both unions functioned like rings, linking on one side French interests and on the other Prussian/Austro-German ones. But while in the early fifties[30] Belgium was the prime mover, Switzerland took over the role as diplomatic leader later in the decade[31].

As emerged from the declarations of the participants in Paris 1855, WETU wanted to open up to all European countries and therefore AGTU members, too[32]. Nowadays various authors attribute France with the undisputed leadership of telegraphic multilateralism in the fifties and sixties, citing as proof, for example, the Paris venue both for the first mixed conference and the WETU and TU founding conferences[33].

---

[29] The expression "imbroglio diplomatique" is borrowed from Patrice Durand Barthez, *Union Internationale des Télécommunications* (Thèse pour le doctorat en droit, Université de Paris I – Panthéon – Sorbonne Sciences Économiques – Sciences Humaines – Sciences Juridiques, 1979), 33.

[30] Bart Van der Herten, Pascal Verhoest, "La contribution belge à la création des réseaux internationaux de communication du XIX siècle," in *Les réseaux européens transnationaux XIXe-XXe siècle: quels enjeux?*, eds. Michelle Merger, Albert Carreras and Andrea Giuntini (Nantes : Ouest Éditions, 1995), 33-47.

[31] Switzerland's role in the fusion of the two unions is covered in detail in Gabriele Balbi, Spartaco Calvo, Simone Fari, Giuseppe Richeri, "Bringing together the two large electric currents that divided Europe. Switzerland's role in promoting the creation of a common European telegraph space (1849-1865)", *ICON* 15 (2009), 61-80.

[32] «La Convention télégraphique de Paris, qui comprend l'échange de 5 États, étant un grand pas vers l'uniformité dans les relations télégraphiques de toute l'Europe», in FF, vol. 1, num. 4, 9 janvier 1856, p. 114. The conference minutes are in AF, Fond E 52, Archive-Nr. 440, Band nr. 2.

[33] Edgar Saveney, "La télégraphie internationale", *Revue des deux mondes* (15 septembre 1872), 360-384 and Leonard Laborie, *La France, l'Europe et l'ordre international des communications (1865-1959)* (Thèse pour obtenir le grade de Docteur, Paris : Université Paris IV – Sorbonne, 2006).

However, recent studies fully supported by primary sources take up a different position, and point to Switzerland as playing an ace diplomatic role right from the beginning[34]. It was, for example, Swiss neutrality rather than formal French leadership which determined AGTU countries being invited to the WETU Conference in Turin, 1857[35]. Austria did not take part because of the rising tension with Piedmont (the Second Italian War of Independence would break out within two years)[36], thereby forcing a change in strategy. The decision was taken to site the following conference at Berne, only after holding some months earlier the Brussels mixed meeting involving Belgium, France and Prussia (June 1858). It can be deduced, therefore, that once Swiss diplomacy had realized it was impossible to get Austria to take part directly in a wider union, it decided to go ahead with a process of standardising the norms of the existing two. The Berne Convention was therefore almost a carbon copy of Brussels and in order to ratify conditions leading to a substantial standardization of telegraph regulations, Switzerland organized yet another conference at Friedrichshafen with AGTU members[37].

So, by 1858 a European telegraph space had been effectively organized, though without any formal consecration[38]. There were however two further obstacles to be overcome before the Unions could come together. The first was that Piedmont and Austria were enemies in the Italian Wars of Independence, while the second was the pressure applied by the WETU countries to adopt a more economical tariff system[39]. Tariffs were, indeed,

---

[34] Gabriele Balbi, *Network Neutrality. Switzerland's role in the genesis of the Telegraph Union, 1855-1875* (Berne: Peter Lang, 2014).

[35] *Légation de S.M. Le Roi de Sardaigne en Suisse à Monsieur Fornerod président de la Confédération*, Berne 30 novembre 1856, in AF, Fond E 52, Archive. Nr. 441

[36] The Swiss Post Department had a clear idea of why the negotiations failed: «Nous avons bien de croire que les raisons qui ont empêché l'Autriche d'accéder à l'invitation de la Sardaigne étaient d'une nature essentiellement politique et resalaient de la situation tendue qui existait entre les deux pays», *Département des Postes et des Travaux publics de la Confédération Suisse au Conseil fédéral Suisse*, Berne 17 août 1857, in AF, Fond E 52, Archive. Nr. 443, Band nr. 2.

[37] Balbi, Calvo, Fari and Richeri, *Bringing together*, 70-75.

[38] Explicit here the words of the *Journal Télégraphique*: «La Convention de Paris [1865] n'a donc fait que consacrer dans la forme et avec un grand progrès de rédaction ce qui avait été fondé en pratique pour les traités de 1858». They refer to the Brussels and Berne Conventions, which both took place in 1858, in the footnote to the article "Le traité de l'Union générale des postes et la Convention internationale des télégraphes", in *Journal Télégraphique,* 35(1874), 575-580.

[39] Van der Herten, Verhoest, *La contribution belge*, 45 ; Saveney, *La télégraphie internationale*, 365

to remain a delicate subject even after the Paris Convention, while the problem of Austria's isolation was resolved diplomatically by Switzerland, which intervened directly with France to get Austria invited to the 1865 Paris event[40]. Austria's participation in turn opened the doors of the Union to Turkey and Russia, so allowing the network to expand eastwards, which was to prove essential for future links with the possessions in Asia. So the 1865 Paris Conference brought the journey to an end by establishing the formal constitution of a telegraph union of European states.

1.2 New container, old contents

On a structural level the TU was profoundly different from its two predecessors because it was born as a single organization mandated to deal with the international telegraph service. On a theoretical level at least, the scale of the new union was to be worldwide, because it was born with the aim of including all the countries desiring to join without any form of discrimination. Substantially then, thanks mainly to the post 1865 presence of Russia and Turkey, the TU embraced a far wider network than AGTU and WETU. But, though the container was indeed much bigger and with an immense potential for expansion, its organizational structures and regulations were tied to what had been put into practice over the previous twenty years.

From an organizational point of view, the TU was mainly based on periodic bodies, i.e. the plenary meetings of representatives from all member countries. During these assemblies delegates were called upon to discuss and establish the basic regulations for the international circulation of telegrams and elaborate documents known as "conventions". The latter had the legal value of international treaties and therefore could only be modified in a new conference. This was why the conventions established general norms and codified details of rules and charges in documents entitled "Regulations" and "Tariff Tables", drawn up exclusively by expert hands and therefore open to modification without a new assembly having to be called. Telegraph agreements had in fact all been international treaties since Vienna 1849, while the conference system had been used by

[40] *Le Département des Postes Naeff au Conseil Fédérale*, 16 juillet 1864, AF, Fond E52, Archive nr. 502 ; *Brunner à Curchod*, 11 août 1864, AF, Fond E52, Archive nr. 502 ; *Curchod à Brunner, lettre confidentiel*, 26 septembre 1864, AF, Fond E52, Archive nr. 502 ; *Conseil Fédéral à Kern*, 18 novembre 1864, AF, Fond E52, Archive nr. 502; *Kern à CF*, 28 novembre 1864, AF, Fond E52, Archive nr. 502; AF, Fond E52, Archive nr. 502, *Brunner à Curchod*, 9 décembre 1864. For a summary of the entire episode see: Gabriele Balbi, *Network Neutrality*, 88-97.

both AGTU and WETU right from their beginnings and the 1857 Stuttgart Conference had provided for the mechanism of rules and tariff tables as attachments to the convention[41].

The norms contained in the Paris Convention were, however, the most apparent affirmation of how the past was taken on board. Only three of the 61 articles, (1, 57 and 61), were completely new. Article 1 stipulated the obligation to adopt special wire in the construction of direct lines in order to guarantee international connections[42] while Article 57 established the obligation to communicate and receive all information about changes in the international network (office hours, line construction/interruption, submarine cable laying/breakages)[43]. The third, article 61, dealt with regulating relations between TU members and the private companies running either land or submarine lines[44]. So, with the exclusion of these three, which, as we shall see, were more challenges for the future than real innovations, international telegraphy was regulated on the basis of well consolidated, familiar norms.

First of all, the Paris Convention took over the five basic principles of international telegraph communication which had been sanctioned by the first Vienna Convention in 1849[45]: 1) freedom of correspondence for all

---

[41] Lagemans, *Recueil des Traités et Conventions conclus pour le Royaume des Pays Bas*, vol. IV, 345-351.

[42] «Les Hautes Parties Contractantes s'engagent à affecter au service télégraphique international des fils spéciaux, en nombre suffisant pour assurer une rapide transmission des dépêches. [...] Les villes entre lesquelles l'échange des correspondances est continu ou très actif seront successivement et autant que possible, reliés par des fils directs, de diamètre supérieur et dont le service demeura dégagé du travail des bureaux intermédiaires», art. 1, de la *Convention Télégraphique Internationale de Paris 1865*.

[43] «Les Hautes Parties contractantes, afin d'assurer, par un échange de communications régulières, la bonne administration de leur service commun, s'engagent à se transmettre réciproquement tous les documents relatifs à leur administration intérieure et à se communiquer tout perfectionnement qu'elles viendraient à y introduire», art. 57, *Convention Télégraphique Internationale de Paris 1865*.

[44] «Les Hautes Parties contractantes s'engagent à imposer, autant que possible, les règles de la présent convention aux compagnies concessionnaires de lignes télégraphiques terrestres ou sous-marines, et à négocier avec les Compagnies existantes une réduction réciproque de tarifs, s'il y a lieu», art. 61, *Convention Télégraphique Internationale de Paris 1865*.

[45] Codding, *The International Telecommunication Union*, 13.

citizens[46]; 2) the right to confidential correspondence[47]; 3) the right of all countries to apply the norms only to international correspondence[48]; 4) the duty to suppress telegrams against public order and morality[49]; 5) the right of every state to suspend international communications when and where deemed necessary[50]. As general principles lay, they answered the dual need of liberal regimes to respect the individual freedom of their citizens and at the same time safeguard national sovereignty.

The same convention had also divided telegrams into three categories: 1) government business; 2) official service; 3) private. Priority went to government telegrams, followed by those on official service and private

---

[46] «Les Hautes Parties contractantes reconnaissent à toutes personnes le droit de correspondre au moyen des télégraphes internationaux», art. 4, *Convention Télégraphique Internationale de Paris 1865.*

[47] «Elles s'engagent à prendre toutes les dispositions nécessaires pour assurer le secret des correspondances et leur bonne expédition», art. 5, *Convention Télégraphique Internationale de Paris 1865*. The subject of the privacy of correspondence is fundamental to all point-to-point communications and marks one of the substantial differences between telecommunications and one-to-many mass media, which are based on the public nature of their communications. See Gabriele Balbi, Juray Kittler, *Dialoguing With Socrates or Disseminating Like Jesus? Rereading Communication History Through 'One-to-One' and 'One-to-Many' Lenses* (presentation, Annual ICA (International Communication Association) Conference, Phoenix, 2012).

[48] «Également animés du désir d'assurer aux correspondances télégraphiques échangées entre leurs États respectifs les avantages d'un tarif simple et réduit, d'améliorer les conditions actuelles de la télégraphie internationale, et d'établir une entente permanente entre leurs États, tout en conservant leur liberté d'action pour les mesures qui n'intéressent point l'ensemble du service», preamble to *Convention Télégraphique Internationale de Paris 1865*.

[49] «Les Hautes Parties contractantes se réservent la faculté d'arrêter la transmission de toute dépêche privée qui paraîtrait dangereuse pour la sécurité de l'État, ou qui serait contraire aux lois du pays, à l'ordre public ou aux bonnes mœurs, à charge d'en avertir immédiatement l'expéditeur», art. 19 *Convention Télégraphique Internationale de Paris 1865*.

[50] «Chaque Gouvernement se réserve aussi la faculté de suspendre le service de la télégraphie internationale pour un temps indéterminé, s'il le juge nécessaire, soit d'une manière générale, soit seulement sur certaines lignes et pour certaines natures de correspondances, à charge d'en aviser immédiatement chacun des autres Gouvernements contractants», art. 20, *Convention Télégraphique Internationale de Paris 1865*.

ones bringing up the rear[51]. The same regulation had already been adopted in all agreements post Vienna (1849)[52].

Lastly, regarding technology, the Paris Conference called for standardization only in terms of adopting the Morse telegraph and code[53], which had already been accepted as the international telegraph by AGTU in 1852 and WETU three years later[54].

1.3 Problems solved, problems to solve

The establishment of WETU in 1855 had put the highlight on two closely linked needs: lower international telegram charges and expand telegraph space. Ten years later the Paris Convention answered both needs with two innovative solutions, the introduction of a uniform telegraph tariff and the inclusion of the Russian and Turkish territories.

Up to 1865 telegraph traffic had been regulated by a zone-based tariff, where charges corresponded to the distance between sending and receiving offices. Clearly copied from the postal service and often used for national tariffs, the system posed some evident complications at international level, particularly over widespread territories like Austria, Russia and Turkey, not to mention the Asian colonies. Though the 1855 Paris Convention managed to attenuate the problems, it did not offer solutions, in that it provided for the reduction of zone tariffs and their expansion, but did not eliminate the injustices inherent in the system. The problem was solved, instead, in Paris 1865, with the introduction of uniform tariffs. Each member state proposed two charges, one for the transit of telegrams over its territory and one for telegrams it sent and/or received.[55] The charge for an international telegram was to be the sum of the dues requested in the sending and destination countries and in all territories the telegram transited. The zone-based tariffs were abolished, except for Russia and Turkey, whose territories were so vast as to warrant having both a European and Asian zone[56].

The continual expansion of the international network had been one of the objectives of the pre-1865 telegraph agreements. Nevertheless, the

---

[51] Art. 7 & art. 11, *Convention Télégraphique Internationale de Paris 1865.*

[52] Codding, *The International Telecommunication Union,* 13.

[53] «L'appareil *Morse* reste provisoirement adopté pour le service des fils internationaux», art. 3. *Convention Télégraphique Internationale de Paris 1865.*

[54] Codding, *The International Telecommunication Union*, 13-20.

[55] Art. 30 & art. 31, *Convention Télégraphique Internationale de Paris 1865.*

[56] The rules for applying tariffs were inserted in articles 32-37, *Convention Télégraphique Internationale de Paris 1865.*

intention to create a common telegraph space was made explicit for the first time during Paris 1855. The key moment actually came with the diplomatic incident of the invitation to Austria. Initially, the conference organizers had decided to invite only the states that had taken part in the 1858 Brussels and Berne Conferences, while in the official invitation it was explained that the conference would serve to uniform definitively the norms established separately the same year[57]. Since Austria had been absent from Berne and Brussels, a diplomatic incident was in the offing. France justified its actions by the fact that the Prussian delegates had been present as representatives of AGTU and therefore indirectly of Austria, too, so that the invitation to Prussia could be interpreted as an indirect invitation to all AGTU members. The Austrians were obviously not at all pleased at being excluded and resorted to a trusted channel in order to express their displeasure[58]. Brunner, head of Austrian Telegraphs, had previously been in charge of Swiss Telegraphs and happened to be on excellent terms with his successor, Louis Curchod. With the official backing of his government[59], Curchod sent a letter to the organizers in which he explained very convincingly the reasons why because of the preceding Braganza and Friedrichshafen treaties, France would have to send a direct invitation to Austria and the other AGTU states not present either at Berne or Brussels[60]. In his long letter Curchod went well beyond the simple defence of Austria to ask the rhetorical question:

> If we want to arrive as soon as possible at drawing up the principles which will be inserted in the general convention for the whole European network

[57] *Ministère de l'intérieur français à le directeur central des télégraphes suisses Curchod*, 12 juillet 1864, AF, Fond E52, Archive nr. 502.
[58] *Brunner à Curchod*, 11 août 1864, AF, Fond E52, Archive nr. 502.
[59] «nous sommes décidé à imposer da nouveau a l'Administration française (ce qui j'avoies déjà eu l'occasion de faire verbalement) la fait de l'existence de la convention de Friedrichshafen, risp. de Bregenz, réunissant à côté des groupes de Berne et de Bruxelles un 3ème groupe d'États européens, qui devrait selon nous être également convoqués à Paris si l'on voulait avoir quelque assurance d'atteindre le but que l'Admin. française sait proposer [...]. Nous avons dit enfin que nous saurons prêts, si la paraissait nécessaire, à provoquer une demande officielle du Conseil féd. au Gouvernement français tendant a obtenir que les États signataires de la convention de Friedrichshafen (Bregenz) soient aussi convoqués aux conférences de Paris», *Curchod à Brunner, lettre confidentiel*, 26 septembre 1864, AF, Fond E52, Archive nr. 502.
[60] *Curchod à de Vougny*, 26 septembre 1864, AF, Fond E52, Archive nr. 502.

and its issues, would it not be better to invite to the conferences all the states in Eastern Europe?[61]

The Swiss, who also put on pressure via their Paris ambassador[62], obtained the considerable result of convincing the head of French Telegraphs to extend the invitation not only to Austria but also to Russia and Turkey[63]. Now the TU would be regulating a network stretching all over Europe, as far as its confines with Africa and Asia. A new state of affairs for the telegraph service and Switzerland's first major success in bringing about international cooperation.

While two important goals were reached, three other innovations at Paris 1865 all posed challenges for the future. Articles 1, 57 and 61 concerned functional aspects of the TU which were as essential as they were contradictory: 1) technical standardization; 2) the need for a permanent body; 3) relations with private companies. All three cases only gave rise to exploratory manoeuvres because the norms in themselves were intrinsically weak. Article 1 provided for the TU members to commit themselves "as far as possible" to laying wider diameter wires along international lines in order to guarantee a rapid and sure service[64]. In the convention project, together with article 3, the only other one to promote technical standardization, it had been mandatory, though the opposition of many delegates had it changed into an option[65]. Article 57 regulated the exchange of reciprocal communications about all modifications to the network[66]. The intermediary was to have been the country which had

---

61 «si l'on veut arriver le plus tôt possible à étendre les principes qui seront posés dans la convention générale à tout le réseau télégraphique européen et à ses aboutissants, ne conviendrait-il pas aussi de convoquer aux conférences les États de l'Extrême Orient de l'Europe?», ibid.

62 *Conseil Fédéral à Kern*, 18 novembre 1864, AF, Fond E52, Archive nr. 502.

63 *De Vougny à Curchod*, 18 octobre 1864, AF, Fond E52, Archive nr. 502.

64 «Les Hautes Parties Contractantes s'engagent à affecter au service télégraphique international des fils spéciaux, en nombre suffisant pour assurer une rapide transmission des dépêches. [...] Les villes entre lesquelles l'échange des correspondances est continu ou très actif seront successivement et autant que possible, reliés par des fils directs, de diamètre supérieur et dont le service demeura dégagée du travail des bureaux intermédiaires», art. 1, *Convention Télégraphique Internationale de Paris 1865*.

65 *Verbaux de la 1ère Séance de la Commission des délégués spéciaux 4 mars 1865*, in *Documents de la Conférence Télégraphique Internationale de Paris*, 115.

66 «Les Hautes Parties contractantes, afin d'assurer, par un exchange de communications régulières, la bonne administration de leur service commun, s'engagent à se transmettre réciproquement tous les documents relatifs à leur

organized the most recent conference, but the difficulties arising in the period 1865-1868 again brought to the fore the possibility of flanking the conferences with a permanent body[67]. Lastly, article 61 proposed extending the convention norms to private companies working on members' territories, guaranteeing as compensation a possible tariff reduction[68]. This meant private companies being accepted into the TU only in a secondary capacity and having to abide by the various governmental decisions. As can be seen, all three norms were in the early stages, underscoring the need to regulate the area but at the same time revealing the TU's functional limits. The member states were so afraid of diminishing their national sovereignty, they were prudent to the point of annulling the effect of the three norms. Though setting off dialogues and discussions surfacing through to the 1920s, technical standardization, the role of a permanent body and relations with private companies were destined to remain highly and heatedly debated issues.

1.4 International to the core

Why did the first international organization devote itself to regulating a telecommunications service? And furthermore, why was the TU established more quickly than the Universal Postal Union (UPU), set up in 1874, and the International Union of Railways (UIC), founded in 1922? The international postal service had indeed been much modernized and expanded in Europe during the thirties and forties[69], and the optic telegraph had spread all over Western Europe, thanks to the French.[70] As

---

administration intérieure et à se communiquer tout perfectionnement qu'elles viendraient à y introduire», art. 57, *Convention Télégraphique Internationale de Paris 1865.*

[67] The French had already proposed in Paris 1865 the creation of a permanent commission. Nevertheless, as in the case of the single tarif, the other delegates found it excessive. *Verbaux de la 9ème Séance de la Commission des délégués spéciaux, 22 mars 1865* in *Documents de la Conférence Télégraphique Internationale de Paris*, 216-217.

[68] «Les Hautes Parties contractantes s'engagent à imposer, autant que possible, les règles de la présent convention aux compagnies concessionnaires de lignes télégraphiques terrestres ou sous-marines, et à négocier avec les Compagnies existantes une réduction réciproque de tarifs, s'il y a lieu», art. 61, *Convention Télégraphique Internationale de Paris 1865.*

[69] Clark, *International Communications*, 15-20.

[70] The optic telegraph network was almost exclusively for government and military use, and the French conquerors imposed it rather than sought cooperation. See Geoffrey Wilson, *The Old Telegraphs* (London: Phillimore, 1976).

these facts show, the electric telegraph was not the first form of communication to expand at an international level during the 19th century. And in fact its first convention (Vienna 1849) had copied both structure and content from the postal agreements of the previous decade[71].

A first reason for its precocity was the low cost of the infrastructures. If compared to railways, roads and optics[72], the telegraph network infrastructure was certainly in no way cost intensive as it consisted of overhead lines made of wooden poles, iron or brass wires and ceramic or glass insulators[73]. A further plus point was that all three elements required fairly simple workmanship and were based on materials like wood and clay, often sourced locally[74]. The postal services then had the downside of added costs, because to cover great distances they had to resort to additional expensive technologies like the railway or steam boat[75].

A second reason came from the telegraph's intrinsically international nature, for like all one-to-one telecommunications technologies, it lent itself easily to communication on an international scale[76]. In the first place,

---

[71] Laborie, *La France, l'Europe*, 150.

[72] The low cost of the electric telegraph in comparison to the optic version was underlined right from the beginnings. For example British newspapers published data showing the notable reduction of fixed costs in the case of state optic telegraphs being replaced by electric ones. See also "Railway travelling, and that noble animal the horse, particularly at this hot season," *The Odd Fellow*, 29 August 1840.

[73] For details on materials and construction see the two internationally used manuals, Tal. P. Shaffner, *The Telegraph Manual. A complete history and description of the semaphoric, electric and magnetic telegraphs of Europe, Asia, Africa and America, ancient and modern* (New York: Pudney & Russel, 1859) ; Robert S. Culley, *A Handbook of Practical Telegraphy* (London: Longmans, Green and Co, 1885).

[74] Obviously this case refers to landlines, which were prevalent, because submarine cables spread from the sixties onward, and relied on very expensive material. For the cost differences between the two technologies see: Simone Fari, *Financing Telegraph Infrastructure (1850-1900)* (Milan: Departmental Working Paper, Department of Economics, Management and Quantitative Methods at Università degli Studi di Milano, 21, 2012).

[75] Alain Gras speaks of the Post as "a technical macro-system of second order" precisely because it needs other technical systems like road, rail or navigation in order to function. Alain Gras, *Le Macro systèmes techniques* (Paris: PUF, 1997), 91.

[76] Stephen D. Krasner, "Communications and National Power: Life on the Pareto Frontier," *World Politics* 43, no 4 (1991): 336-366.

its international nature was connected to its most evident feature: speed[77]. Telegraph transmissions were then far more rapid than any other kind of communication and the advantage increased with the distance covered, thus favouring expansion beyond national borders, which was especially important for small and medium-sized states. In contrast, international postal cooperation lost ground relatively, precisely because of its lack of a rapid means. It was no coincidence in fact that the Universal Postal Union was established only after the construction of important rail connections[78].

A good postal service could compete on equal terms with the telegraph over short distances, both for speedy delivery and prices. As long as the message was to be sent only a few hundred kilometres, the telegraph was not much faster than the post, though with higher costs and less reliability[79]. Where the telegraph was revolutionary was over distances of more than a thousand kilometres, which only happened with international connections beyond Europe. With the exclusion of Austria, Russia and Turkey, the geography and history of Europe had led to the formation of small states, and these consequently needed to activate international lines right from the beginning in order to exploit the potential on offer. Thus international telegraphy was developed at the same time as the domestic service, and not, as could be expected, later[80].

The international emphasis on the telegraph must be linked to another important feature of one-to-one telecommunications: coverage, i.e. the number of participants in the network. The telegraph had indeed generated coverage from the beginning both at home and abroad[81]. Some hold that the efficiency of communication networks increases automatically with the

---

[77] «Speed is the most striking characteristic of telecommunications», Daniel H. Headrick, *The Invisible Weapon. Telecommunications and International Politics 1851-1945* (New York, Oxford University Press, 1991), 4.

[78] Van der Herten, Verhoest, *La contribution*, 37.

[79] With telegraphs the message had to be coded, sent and decoded, therefore increasing the possibility of errors in transmission. This meant reliability was a quite subjective criterion, so that all things equal, clients already using one service preferred to avoid experimenting a new one.

[80] «En Europe, les communications à longue distance demandaient un degré de coopération intergouvernemental sans précédent dans l'histoire, à cause du nombre de frontières à travers les quelles devaient être acheminés les messages», James Foreman-Peck, "L'État et le développement du réseau de télécommunications en Europe à ses débuts," *Histoire, économie et société* 3, no 8 (1989) : 384.

[81] «*Coverage* refers to the number of participants in a network. The telegraph is not an isolated device, but appeared from the very beginning as a network, that is to say a complex socio-technical system spread over a large geographic area». Headrick, *The Invisible Weapon*, 4.

number of participants, generating what is termed a "network effect" or "network externality"[82]. In Europe this phenomenon was able to generate only thanks to international cooperation, in that the small dimensions of the single states prevented them from reaching a sufficient scale to produce network effects by themselves[83].

Finally, the international nature of the telegraph is to be attributed to the kind of service offered, and here again a comparison with the Post is eloquent. The latter had been conceived of as a universal service, with as potential users all those able to read and write. The telegraph service was born instead at a domestic level (in Great Britain and the United States), as an exclusivity for businessmen, entrepreneurs and brokers needing rapid, frequent communications[84]. However, leaving aside Great Britain, in Europe the international service was born under the auspices of single governments. The order of priority which privileged government telegrams over all others shows that from 1849 onwards international telegraphy answered first of all a diplomatic need. So while the Post was conceived of as basically national in the forties and fifties, the telegraph was already at the service of diplomacy and the business world and had a greater propensity for international connections[85].

---

[82] Albert-László Barabasi, *Linked: The New Science of Networks* (Cambridge Massachusetts: Perseus Publishing, 2002).

[83] Van der Herten, Verhoest, *La contribution*, 44.

[84] «The fact is, that Postal correspondence and Telegraphic correspondence are two very distinct things. They are NOT "in the same category", or in anything approaching the same category. A broker in London wants to supply his correspondents in the country with prices and information of various descriptions, on which the success of their daily operations is dependent. He resorts to the instantaneous communication afforded by the Telegraphic wire. On the other hand, a lady in London desires to afford to her country cousins and acquaintances the fullest information upon household, domestic, and family affairs of interest, the particulars of which require to be conveyed in full, but which are not of pressing or immediate importance. She resorts to the cheaper, but less expeditious, machinery of the Post. The Telegraph and the Post have thus their respective duties to perform: each, no doubt equally important to the public, and each, in their respective ways, of public interest. But the two modes of communication are not "in the same category". The two things are perfectly distinct and separate». *Government and the Telegraphs. Statement of the case of the Electric & International Telegraph Company against The Government Bill for acquiring the Telegraphs* (London: Effingham Wilson, 1868), 53.

[85] Van der Herten, Verhoest, *La contribution*, 37.

1.5 Politics and Eurocentrism

The twenty-year journey which led to the establishment of the TU was dominated by political interests and intense diplomatic activity[86]. All the treatises from the first conventions to the creation of the TU were the results of the activity carried out by delegations composed of diplomats and high-ranking technicians. While the former were state officials with powers of representation, the latter were mostly the heads/general inspectors of national telegraph administrations. Up to the mid-fifties, therefore, international regulation came exclusively from the directives of national governments and the technical expertise of top telegraph men[87]. In the same way, up to 1871, there was neither the direct or indirect participation of private capital, because the conventions only concerned states and automatically excluded any privately-run services. Then from 1855, on the rare occasions in which private companies were invited, they had neither right of vote or consultation. The most evident proof of the pre-1865 domination of politics is given by the basic dynamics guiding the various operations. The first agreements signed in 1848-1849 by Prussia and the German states fitted in perfectly with the journey towards German unification. More generally, the establishment and development of AGTU mirrored the change in equilibrium in the German-speaking area, where Austria, the traditional power figure, found itself having to come to terms with Prussia's ambition and dynamism. The foundation of WETO and France's constant desire to be in the limelight (Paris 1855 and Paris 1865) reflected its worry over the need to curb Prussia's political thrust. Furthermore, the role of Switzerland, which, as we have seen, was fundamental in the TU's expansion to the east, was not limited to telegraphy alone, but gave the first signs of the country operating as an international prime mover. From the mid-1850s onwards in fact Switzerland carried out an ever-increasing diplomatic role at the centre of the European chessboard, always playing as its trump card its much vaunted neutrality[88].

---

[86] This is well explained in Verdiana Grossi, "Technologie et diplomatie suisse au siècle: le cas des télégraphes," *Relations Internationales* 39 (automne 1984): 287-307.

[87] Laborie gives a detailed description of the diplomats and technicians taking part in the post and telegraph conferences in the second half of the nineteenth century. Laborie, *La France, l'Europe*, 266-272.

[88] For an overview of the history of Swiss neutrality: Hans Ulrico José, "Origines, interprétations et usages de la 'neutralité helvétique'," *Matériaux pour l'histoire de notre temps* 93 (2009), 5-12 ; Marko Milivojevic, Pierre Maurer (eds.), *Swiss*

The dynamics underlying the gestation process of the TU do not only prove the overpowering presence of politics, but also emphasize the important role of continental Europe. The exclusion of Great Britain and the absence of Russia and Turkey up to 1865 contributed to making the Eurocentrism more continental and western. This can be explained by contextualizing the role of Europe in the international politics of the mid-century. The Old Continent was still the centre of political and economic power, the United States was still a work in progress, while Asia and Africa were undergoing the ever-encroaching ambitions of Europe[89]. In this context international cooperation could only be conceived of as European.

It is hardly surprising, therefore, that international cooperation was understood as existing among European states and the adjective "European" was used for international telegraph organizations with a meaning closer to "world" as we use it today[90]. In WETU, for example, it was declared during the 1855 founding conference in Paris, that there "will be later admitted all the European states which are desirous of joining"[91].

---

neutrality and security : armed forces, national defence and foreign policy (New York: St. Martin's Press, 1990); Ann G. Imlah, *Britain and Switzerland 1845-60. A study of Anglo-Swiss relations during some critical years for Swiss neutrality* (Hamden (Connecticut): Archon Books, 1966); Max Liniger-Goumaz, *La Suisse, sa neutralité et l'Europe* (Genève : Éditions du Temps, 1964); Georges Perrin, *La neutralité permanente de la Suisse et les organisations internationales* (Heule: UGA, 1964).

[89] Christopher A. Bayly, *The Birth of the Modern World 1780-1914. Global Connections and Comparisons* (Malden, Oxford, Victoria : Blackwell Publishing, 2004).

[90] Significant the terminology used by a reporter of the times: «On ne lira pas sans fruit l'histoire des efforts qui ont été faits pour fonder en Europe une véritable union télégraphique», «on comprendra que le service européen devait être rapidement entravé par une confusion croissante, et que l'on devait avoir dès lors l'idée de réunir toute l'Europe dans une convention unique», «la France s'efforça d'en introduire le principe dans le service européen», «voilà donc les représentants de l'Europe entière assemblés pour régler un grand objet d'utilité commune», «c'est une œuvre fructueuse entre toutes et véritablement digne d'intérêt que celle que les délégués de l'Europe viennent de faire sous nos yeux autour du tapis vert de la conférence de Paris. Quelles que soient les questions traitées, n'est-ce point un spectacle plein d'enseignements que de voir les représentants de l'Europe entière assemblés dans un dessein d'utilité commune ?», Saveney, *La télégraphie internationale*, 359-384.

[91]«seront admis successivement tous les États de l'Europe qui voudront en faire partie», Séance de la conférence internationale de Paris, 5 décembre 1855, in AF, Fond E 52, Archive-Nr. 440, Band nr. 2.

In the same way, at the opening of Paris 1865, the French minister announced that "France felt it necessary to suggest not only to all the signees of the preceding conferences but all European powers to call a conference to negotiate a general treaty"[92].

From this point of view, telegraphic multilateralism and more specifically the management of the TU are significant elements in the thesis which holds that European integration went through a formative moment during the nineteenth century by means of the tangible construction of numerous infrastructural networks[93]. In this process the TU is said to have favoured the formation and consolidation of a continental telegraph network which was an essential contribution towards the creation of a "European dimension of technology"[94].

According to Misa and Shot, the following concepts are essential: *linking*, referring to connections realized; *circulation*, for movement of persons, objects and information; *appropriation*, for the personal use made of the means of communication[95]. In other words, the so-called tensions about international cooperation between 1848 and 1865 in creating telegraph connections between the principal European cities (*linking*), standardizing the flow of information (*circulation*) and using telegraphs for business and personal reasons (*appropriation*) were very early examples of the continent's tangible integration[96].

1.6 Liberalism

As is well known, Hobsbawm defines the period 1848-1870 as the age of capitalism and identifies it as the triumph of liberal society[97]. The process of the TU's gestation, which belongs to the period, was indeed

---

[92] «la France a cru devoir proposer non seulement aux États signataires des Conventions précédents, mais à toutes les Puissances de l'Europe, de former une Conférence appelée à négocier un Traité général» *Séances Générales, première séance, 1er Mars 1865*, in *Documents de la Conférence Télégraphique Internationale de Paris*, 76.

[93] Frank Shipper, Johan Schot, "Infrastructural Europeanism, or the Project of building Europe on infrastructures: an introduction," *History and Technology* 27, no 3 (2011): 245-264.

[94] Ibid.

[95] Thomas J. Misa, Johan Schot, "Inventing Europe. Technology and the Hidden Integration of Europe," *History and Technology* 21, no 1 (2005): 1-20.

[96] http: www.tensionsofeurope.eu, accessed 5/4/2013.

[97] Eric Hobsbawm, *Age of Capital 1848-1875* (London: Hachette, 2010).

much imbued with liberal thought and beliefs[98]. The principles of the free circulation of telegrams, the right of all citizens to send them plus the state's duty to suppress any that could undermine it politically or morally, were inserted in all the telegraph conventions from 1850 onwards. In addition, the states promoted the circulation of information by progressively reducing basic charges and gradually extending the network.

The fact that politics dominated the whole process and that most states ran the service directly was not in contrast with the tenets of liberalism. Government decisions spelled politics and with few exceptions, the governments of the period sought to confirm them by referring to liberalism, or were actually liberal in nature. Belgium, for example, which led the attempt at the fusion of the two Unions in the early fifties, was the only European country where a liberal revolution had managed to triumph during the 1830-31 uprisings[99]. In its turn, Switzerland's federal government, which so skilfully helped guide the establishment of the TU, was the only liberal government to come through from the 1848 revolts which had so shaken the whole of continental Europe to the core[100]. Furthermore, as some liberal economists had well illustrated during the 1869 nationalization of the British telegraphs, a public monopoly in this field was not in contradiction with the principles of economic liberalism[101]. As in the case of the Post, the telegraph would function as a basic infrastructure for the free circulation of information, indispensable in its turn as a guarantee of the free market[102]. Finally, as Hobsbawm

---

[98] On the relations between the international telecommunications system and nineteenth century capitalism see: Dwayne R. Winseck, Robert M. Pike, "Communication and Empire. Media markets, power and globalization, 1860-1910," *Global Media and Communication* 4, no 7 (2008): 7-36.

[99] Els Witte, *Political History of Belgium from 1830 Onwards* (Bruxelles: VUB Press, 2000).

[100] Pierre du Bois, *Le Guerre du Sonderbund. La Suisse de 1847* (Paris: Éditions Alvik, 2002) ; Roland Ruffieux, "La Svizzera dei Radicali", in *Nuova Storia della Svizzera e degli Svizzeri* (Bellinzona: Edizioni Casagrande, 1983).

[101] William Stanley Jevons, "On the Analogy between the Post Office, Telegraphs, and other Systems of Conveyance of the United Kingdom, as regards Government Control", in William Stanley Jevons, *Methods of Social Reforms and other papers* (London: MacMillan and Co., 1883), 279-280.

[102] "Moreover, if one examines Ricardo's analysis of the shipping industry under the Navigation acts and his description of the telegraphs before nationalisation, the parallels are remarkable. In each an oligopolistic system caused unnecessarily high prices and inhibited the expansion of the economy. Only his solutions differed. In the case of the shipping industry, free trade was the remedy; for the telegraphs, nationalisation", Charles Richard Perry, *The Victorian Post Office. The growth of a*

highlights, the nationalistic impulse of the single states, which would find expression in the following age of imperialism (1870-1914), sought its coherence in the international cooperation of liberalism[103]. The TU, a full expression of its tenets, limited national sovereignty only insofar as it guaranteed an indispensable service for the circulation and materialization of liberal principles, of which the single national governments were indeed the expression.

1.7 Conclusion

As we have seen, the structure and the norms of the TU were not born in Paris in 1865, but were the fruit of a twenty-year long gestation which consolidated certain characteristics destined to last beyond WWI.

First of all, the fact that the norms approved in Paris were mainly the result of innumerable preceding treatises confirmed a conservative tendency that would often appear in the defence of consolidated rules. Secondly, as years went by the initial interpretation of the telegraph as an international service favoured greater differences in management and tariffs between domestic and international services, with consequent effects on the concept of service, either as exclusive or universal. Thirdly, the pre-1865 invasive role of politics was sanctified and officialised by the organizational structure of the TU, which led to a promotion of the roles of government delegations and the exclusion of private bodies. Finally, the liberalism already influential in the gestation period was absorbed as a political philosophy, and with its national variations and internal contradictions weighed on the formation of the norms.

---

bureaucracy (Wooldbridge: The Royal Historical Society-The Boydell Press, 1992), 80.

[103] Hobsbawm, *Age of Capital*, Chapter 6, Section IV.

CHAPTER TWO

THE CONFERENCES

The conferences were quite simply the meetings of those representing the members of the TU. However, the situation is not so simple when we try to define what was meant by a member and a representative. Who could belong to the TU? States? Telegraph administrations? Private companies? And once we have defined what a member of the TU was, who could act as a representative? A diplomat? A top telegraph manager? A company president?

Sections 1, 2 and 3 describe the main players in the conferences - governments, telegraph administrations and private companies. There then follows an account of those who physically took part in the assemblies, i.e. who represented the member states (Section 4), together with a discussion of the conferences' competences and the procedures used to elaborate and approve the TU norms (Section 5).

2.1 States vs telegraph administrations

During our period those classified as effective members of the TU had joined by giving official notice of adhesion and signing the convention, the main document drawn up at each conference. Since the convention followed pre-TU customs and had the standing of an international treaty, the Paris Convention was signed by the diplomatic delegates from all states present and then ratified by the national legislative bodies. The convention's status as an international treaty was contained in article 56, which provided for it to be reviewed by the diplomatic representatives of all the states meeting periodically by means of an international conference[1]. The article was left unaltered in the following conventions of

[1] «La présente Convention sera soumise à des révisions périodiques, où toutes les Puissances qui y ont pris part seront représentées.», art. 56, *Convention télégraphique internationale de Paris 1865.*

Vienna 1868 [2] and Rome 1871[3], while during the 1875 St. Petersburg Conference the decision was taken to approve a definitive version. The result was that post 1875 no modifications were made to the Convention's text, and only dispositions of an administrative nature such as tariffs and regulations were implemented. It also followed that those taking part were now the administrative representatives of the member states, and had none of the plenipotentiary powers of diplomats[4].

Introduced in Paris, the process of adhesion provided for an aspirant member to make application to the country which had organized the most recent conference, and notice of the new adhesion would then be given to all members[5]. Basically in the period from 1865 to post WWI, the members were all states with government-run telegraphs. As Great Britain and the States had private services, neither was present either at Paris or Vienna and it was only after the nationalisation of its overhead telegraphs in 1869 that Great Britain joined the TU, while the States always remained outside[6].

---

[2] «La présente Convention sera soumise à des révisions périodiques, où toutes les Puissances qui y ont pris part seront représentées.», art. 62, *Convention télégraphique internationale de Vienne 1868.*

[3] «La présente Convention sera soumise à des révisions périodiques, où toutes les Puissances qui y ont pris part seront représentées.», art. 61, *Convention télégraphique internationale de Rome 1872.*

[4] «Le tarif et le règlement prévus par les articles 10 et 13 sont annexés à la présente Convention. Ils ont la même valeur et entrent en vigueur en même temps qu'elle. Ils seront soumis à des révisions où tous les États qui ont pris part pourront se faire représenter. A cet effet, des Conférences administratives auront lieu périodiquement, chaque Conférence fixant elle-même le lieu et l'époque de la réunion suivante» ; «ces Conférences sont composées des délégués représentant les Administrations des États contractants». Art. 15 & art. 16, *Convention télégraphique internationale de St. Pétersbourg 1875.*

[5] «Les États qui n'ont point pris part à la présente Convention seront admis à y adhérer sur leur demande. Cette adhésion sera notifiée par la voie diplomatique à celui des États contractants au sein duquel la dernière conférence aura été tenue, et, par cet État, à tous les autres. Elle emportera, de plein droit, accession à toutes les clauses et admission à tous les avantages stipulés par la présente Convention», art. 60 *Convention télégraphique internationale de Paris 1865.* The same norm is repeated in art. 65, *Convention télégraphique internationale de Vienne 1868*, art. 63, *Convention télégraphique internationale de Rome 1872* and art. 18, *Convention télégraphique internationale de St. Pétersbourg 1875.*

[6] Actually, in the late 1870s the TU showed both formal and informal signs of opening towards the possible membership of the States. First of all art. 64 of the Rome Convention provided for private companies servicing one or more states connected to the international network to be considered part and parcel of the

There are those in literature who maintain that sovereign states were never real members of the TU. George Codding, for example, holds there are numerous elements pointing to the real members of the Union being the individual telegraph administrations[7], which he sees as "any department or service of a government responsible for the functioning of a telegraph service"[8]. Backing this idea there is also the fact that the telegraph administration delegates alone played an active role in the conference proceedings, and after the creation of the administrative conferences were indeed the only participants, as set down in article 16 of the St Petersburg Convention[9]. However, Codding's reading can also be easily refuted, as Patrice Durand Barthez does[10], by simply quoting directly from the St Petersburg Convention terms like: «Les Hautes Parties contractantes»[11], «Chaque Gouvernement»[12] and «des États contractantes»[13]. What emerges clearly is that although the diplomats carried out only a formal function,

---

state's or states' national network. This meant that states with privately-run telegraphs could join the Union. «Les exploitations télégraphiques privées, qui fonctionnent dans les limites d'un ou de plusieurs États contractants avec participation au service international, sont considérées, au point de vue de ce service, comme faisant partie intégrante du réseau télégraphique de ces États», art. 64, *Convention télégraphique internationale de Rome 1872*. There was also an intense diplomatic activity carried out by the Bureau and the organizers of the Rome and St Petersburg conferences to get delegates from the States. Before the Rome Conference Curchod as head of the Bureau sent a letter to Ernesto D'Amico, the head of Italian Telegraphs, recommending him to do everything possible to welcome conference delegates from private American companies and/or government representatives. He detailed: «Je ne saurais donc qu'approuver l'intervention du gouvernement des États Unis, s'il est, lui, disposé à prendre en main les intérêts des compagnies télégraphiques américaines, et si celles ci, de leur côté, acceptent d'être représentés de cette façon-là», ITU-Corr, feuille n. 85/4, 1869. Despite all the efforts, the States decided to stay out of the TU, though they did send their ambassador in Russia as an observer to St Petersburg. *Documents de la Conférence Télégraphique Internationale de St Petersburg* (Berne: Bureau International des Administrations Télégraphiques, 1875), 395.

[7] Codding Jr., *The International Telecommunication Union*, 38.

[8] Ibid., note 166.

[9] «Ces Conférences sont composées des délégués représentant les Administration des États contractants», art. 16, *Convention télégraphique internationale de St. Pétersbourg 1875*.

[10] Durand Barthez, *Union Internationale des Télécommunications*, 56.

[11] Art. 1, art. 7, art. 9, art. 12 & art. 17, *Convention télégraphique internationale de St. Pétersbourg 1875*.

[12] Ibid., art. 4 & art. 8.

[13] Ibid., art. 11, art. 13 & art. 14.

the telegraph conventions were in both form and substance international treaties, and therefore drawn up by sovereign states. The same was also true for the St Petersburg Convention, the benchmark for the rest of our period.

The debate on the identity of TU members has not involved only later scholars, but also emerged in the period itself, particularly at the Rome Conference. For example, the private submarine company representative Jules Despecher[14] put forward an amendment to article 13 of the convention which provided for the governments to choose the route for a telegraph if there was no sender specification[15] and suggested replacing the word "governments" with "administrations"[16]. Though the proposed change may have seemed purely formal, it was substantial for two reasons: 1) an administration could always choose the technically superior way regardless of political calculations; 2) the companies could see themselves as administrations and could well opt for the most profitable route.

The choice of one or another line could well determine a greater or lesser income for the sender administration while at the same time penalizing intermediary offices. Beyond the question of article 13 and its outcome, the debate was extremely interesting because it highlighted the fact that delegates had different ideas of what a TU member was. During the debate, three lines of thought emerged, with some seeing its identity in terms that were: 1) governmental; 2) administrative; 3) intermediary/ technical. The German delegate Meydam spoke for the governmental line and upheld with vigour the priority of politics:

> Government choices over routes are not always determined by considerations of service. There can also be political reasons which lead them to choose this or that direction because of the different kinds of

[14] Despecher represented seven submarine telegraph companies : 1) Falmouth, Gibraltar and Malta Telegraph; 2) Marseille, Algiers and Malta Telegraph; 3) Anglo-Mediterranean Telegraph, 4) British Indian Submarine Telegraph; 5) British Indian Extension Telegraph; 6) British Australian Telegraph; 7) China Submarine Telegraph. He also represented the company "Construction and Maintenance Telegraph". *Documents de la Conférence Télégraphique Internationale de Rome* (Berne : Bureau International des Administrations Télégraphiques, 1872), 261.

[15] «Chaque Gouvernement reste juge, vis-à-vis de l'expéditeur, de la Direction qu'il convent de donner aux dépêches, tant dans le service ordinaire qu'au cas d'interruption ou d'encombrement des voies habituellement suivies», art. 13, *Convention télégraphique internationale de Vienne 1868*.

[16] «Remplacer le mot "Gouvernement par la mot "Administration"», amendement de la Russie et proposition de Sous-marine unies, art. 13, *Documents de la Conférence Télégraphique Internationale de Rome*, 102.

telegrams or correspondence exchanged with certain countries. Administrations would obviously not be competent about such matters. Allowing them to choose the routes would also create other difficulties, for one administration's preferences for this or that line could block the corresponding country's arrangements and create competition, which the spirit of the Convention justly tries to avoid.[17]

The Austro-Hungarian delegate Brunner thought differently, and held that the term administration was more suitable to service needs. He identified two different circumstances in routing, either normal or emergency, i.e. in the case of breakdowns. In the first case the administration would comply with its government's decision:

> In the second case, where an immediate decision has to be taken to keep the service going, the administration and not the government resolves the question. The term "Administration" would therefore seem the best choice for all eventualities[18].

Finally Vinchent, the Belgian master mediator, pointed out that in reality the first interpretation did not exclude the second. He argued that a distinction had to be drawn between the three concrete cases: 1) where the administration was governmental, and therefore followed its government's wishes; 2) where the administration was in the hands of a private company with a contract obliging it to follow government orders; 3) where the administration was in the hands of a completely independent company. He argued that only in the last case would it have any sense to use the word administration and not government. However, he felt that such a definition had only one practical application, and therefore it was not worth while going through the motions and approving Despecher's substitution[19].

---

[17]«Le choix de voies qui appartient aux gouvernements n'est pas toujours déterminé par des considérations de service. Il peut y avoir également des motifs politiques qui amènent à choisir telle ou telle direction pour différentes natures de dépêches ou pour les correspondances échangées avec certains pays. Les Administrations évidemment ne seraient pas compétentes en pareille matière. Leur attribuer le choix des voies aurait, d'ailleurs, d'autres inconvénients ; car les préférences d'une Administration en faveur de telle ou telle ligne pourraient paralyser les dispositions prises par le pays correspondant et créer des concurrences que l'esprit de la Convention tend justement à écarter», ibid., 292.

[18] "Pour le second cas, où il s'agit de décision immédiate prise uniquement en vue de considérations de service, c'est l'Administration et non le Gouvernement qui peut résoudre la question. La terme "Administration" lui paraitrait donc répondre mieux à toutes les éventualités», ibid., 293.

[19] Ibid, p. 294.

At the end, the government and intermediary/technical positions came out on top, which implied the majority tended to still be thinking in government rather than administration terms. The amendment was in fact rejected[20], so showing how delegates held on to the view that the states were the effective members of the TU.

2.2 The colonial vote

Identifying sovereign states or administrations as union members was not merely a question of terminology but had extremely important practical implications. The right to vote in the conferences was strictly linked to a clear-cut definition of what a union member was. Voting rights were granted each time at the opening of a conference and at both Paris and Vienna each member country had the right to one vote alone[21]. In Rome however Great Britain was given two votes, one for the home country and the other for British India[22]. In order to prevent other states with colonial possessions taking excessive advantage of multiple voting, the Russian organizers of St Petersburg 1875 suggested reforming the system[23]. In particular, they proposed allocating extra votes according to the extent of telegraph networks, a provision that would obviously give an

---

[20] In reality, polemics over the amendment did not finish with the vote going against it. The article was then sent on to the committee, which drew up a new text reconciling the different points of view. With a technicality both the terms 'administration' and 'government' were eliminated: «Lorsque l'expéditeur n'a prescrit aucune voie à suivre, chacun des offices à partir desquels les voies se divisent, reste juge de la direction à donner à la dépêche», art. 14, *Convention télégraphique internationale de Rome 1872*. See also *Documents de la Conférence Télégraphique Internationale de Rome*, 294. Interesting the motivation that Curchod, the committee chair, put forward to explain replacing "Governments" with "Offices": «M. Curchod ajoute que la Commission, après avoir étudié les propositions qui lui ont été renvoyées et s'être inspirée de la discussion qui a eu lieu à leur sujet, a voulu faire ressortir les deux cas qui se présentent, selon que la voie est ou n'est pas indiquée par le public. Elle a tenu compte de l'amendement de l'Allemagne, mais en remplaçant dans le premier paragraphe le mot "Gouvernement" par l'expression plus vague d'Office. Afin, d'ailleurs, d'éviter les inconvénients qui auraient pu se produire, si les Compagnies avaient eu la faculté d'apprécier les circonstances dans lesquelles il serait loisible de ne pas se conformer aux intentions de l'expéditeur, la Commission a limité d'une manière générale cette dérogation au cas de l'interruption de la voie désignée», ibid., 565.

[21] Ibid., 263.

[22] Ibid., 263.

[23] ITU-Corr., feuille n. 29/1, 18 décembre 1873.

advantage to very big countries like Russia itself. The proposal was mediated by the Bureau, which submitted it to the opinion of the other administrations[24]. An ample debate followed, from which there emerged both different projects on voting and different ideas on how a TU member should be defined[25]. Generally speaking, those in favour of a multiple vote tended to consider the single telegraph administrations union members and therefore found it fitting to allocate votes in proportion to the size of the networks and the administrators' capacity to run them. Some, like Russia, felt it was right to give votes according to the kilometres of lines[26], while others like Italy would have preferred a criterion which took into account the quantity of international telegrams[27] or, like Spain, called for a complex system of diverse indicators such as the number of offices, the sum of domestic and international telegrams and the extent of the lines[28].

Those asking for a single vote held that only sovereign states could be legitimately considered Union members. Several of them were consequently against the colonies and private companies obtaining member status. The head of Swiss telegraphs held, for example, that:

> We venture to point out about this matter that the International Telegraph Convention is a diplomatic act concluded in a vote in which each state took part with the same number of votes and that consequently it seems to us inadmissible that modifications be introduced to make it different[29].

As in many other cases, a process of mediation between the different opinions took place over the way of voting. Initially, Curchod as head of the Bureau sent a confidential letter to the Russian Administration to give his opinion about any modifications to the voting criterion at the conferences[30]. He held it inappropriate to allocate more than one vote per state, but if the administrations taking part were favourable, the ruling would have to be applied in administrative and not diplomatic conferences. Therefore Curchod's idea was to hold a last diplomatic

---

[24] ITU-Corr., feuille n. 29/2, 10 janvier 1873.

[25] ITU-Corr., dossier n. 29, 1873.

[26] ITU-Corr., feuille n. 29/1, 18 décembre 1873.

[27] ITU-Corr., feuille n. 29/17, 6 juin 1873.

[28] ITU-Corr., feuille n. 29/12, 27 février 1873.

[29] «Nous nous permettons de vous faire observer à ce sujet que la Convention Internationale Télégraphique est un acte diplomatique conclu par une votation à la quelle chaque État représenté participait avec le même nombre de voix el qu'en conséquence il nous semble inadmissible qu'on adopte pour les modifications à y apporter un mode différent» ITU-Corr., feuille n. 29/7, 31 janvier 1873.

[30] ITU-Corr., feuille n. 29/22, 11 juillet 1873.

conference in which it would be decided that following conferences would be only administrative. He held in fact that:

> The aim in giving in the administrative conferences a deliberative voice to each distinct administration is to profit from the experience and understanding of each of them and overall put them in a position to give their views[31].

The conclusion was to abandon the idea of the multiple vote proposed by Italy, Russia and Spain and simply allocate one vote to each administration, while approving a colonial vote. Colonial telegraph administrations were in fact considered separate entities and as such had a right to take an active part in the conferences. This line was followed by the Bureau in drafting the project of the convention[32] and after a thorough discussion it was approved[33] and inserted in the St Petersburg Convention[34].

The issue of the colonial vote and its affirmation post 1875 needs to be interpreted as a sign of the progressive changes in the political balance within the TU. During its first ten years of life the TU went on being just as Eurocentric as before. For example, still in 1875, in the presence of observers from the United States, Brazil and Japan, Ernesto D'Amico, head of Italian Telegraphs and chairman of Rome 1872 felt the need to point to the TU's strongly European identity[35].

---

[31] «le but qu'on se propose en donnant, dans les Conférences administratives, une voix délibérante à chaque administration distincte est de profiter de l'expérience et de lumières de chacune d'elles et surtout de les mettre à même de faire valoir leurs opinions», ITU-Corr., feuille n. 29/25, 28 septembre 1873.

[32] *Project de Convention télégraphique internationale*, in *Documents de la Conférence Télégraphique Internationale de St Petersburg*, 121-128.

[33] The subjects of the correspondence leading up to the drafting of the convention project anticipate the debate. The delegate defended more or less the same positions they had held in their correspondence with the International Bureau. See also ibid., 277-287.

[34] «Ces Conférences sont composées des délégués représentants les Administrations des États contractants. Dans les délibérations, chaque Administration a droit à une voix, sous réserve, s'il agit d'Administrations différentes d'un même Gouvernement, que la demande en ait été faite par voie diplomatique au Gouvernement du pays où doit se réunir la Conférence, avant la date fixée pour son ouverture, et que chacune d'entre elles ait une représentation spéciale et distincte», art. 16, *Convention télégraphique internationale de St Petersburg 1875.*

[35] «Du reste, tous les États européens font désormais partie de l'union, et il est évident que l'adhésion successive des États extra-européens ne pourrait avoir lieu

Whatever, Great Britain's entry in 1872 was destined to modify over time the continental equilibrium of the TU. As the main economic and political power in the period, Great Britain was developing an almost hegemonic global system of telecommunications. While its delegates kept a fairly low profile at Rome, on more than one occasion at St Petersburg they expressed their grievance at being relegated to a relatively marginal position, and called attention to the problem of both the colonial vote and the mainly British financed submarine cable companies. For example, during the approval of lesser norms in the eleventh sitting, the British India delegate Robinson brought up the subject of the continual tariff reduction for India established by the European powers at Vienna and Rome, and claimed that it had caused losses for extra-European administrations[36]. The other British delegate was even more explicit:

> Mr Champain begs the Conference not to forget this very true fact, that the great stretches of extra European telegraphs, which have been of advantage to the whole world, were built at great expense, with considerable risk, without the European administrations contributing a cent to the expenses or running any of the risks. The extra-European Administrations, it is true, have drawn profit from the European lines, but they are lines open to all and no special reduction has been made in the interest of their correspondents. […] Another extremely important fact which concerns the present proposition of the Indian delegates is that Great Britain is without any contradiction the commercial centre of the world and consequently the number of extra-European telegrams arriving or leaving the country is far higher than that of the whole of continental Europe. Perhaps the members of the Conference will be surprised to learn that without counting London, which covers two thirds of the correspondence in Britain, the rest of the extra-European traffic in the country is more than double the traffic of all the European states. Now each telegram from the English provinces, Scotland and Ireland on one side and India on the other costs one franc more than London postage. Furthermore, the other main centres of European business with the Far East are located in the western part of continental Europe.[37]

---

que très-lentement», *Documents de la Conférence Télégraphique Internationale de St Petersburg*, 274.

[36] «Les Administrations européens n'ont en rien supporté leur part de ces charges, et elles ont, au contraire, retiré des bénéfices sensibles provenant de l'accroissement du nombre des télégrammes», ibid, 467.

[37] «M. Champain prie la Conférence de ne pas oublier ce fait rigoureusement vrai que les très grandes étendues des télégraphes extra-européens, dont le monde entier a profité, a été construite à grands frais et avec des risques considérables, sans que les Administrations européens aient supporté un centime de ces dépenses ni couru

Retaliation was soon made to this British attack on the powers at the heart of the TU. The French representative retorted that the traffic towards the Indies was mostly British and therefore the European powers certainly could not entertain losses in order to sustain its development[38]. The Belgian Vinchent followed the line that the European administrations had already made a big concession in approving the introduction of charging by the word for extra-European telegrams[39]. The contrast between British and continental interests continued to come to the surface in the debates that followed.

---

rien de ces risques. Les Administrations extra-européens ont, il est vrai, profité des lignes de l'Europe, mais ce sont des lignes ouvertes à tous et aucune réduction spéciale de taxe n'a jamais été faite dans l'intérêt de leurs correspondances. [...]Un autre fait de la plus haute importance, en ce qui concerne la proposition actuelle des délégués des Indes, c'est que l'Angleterre est sans contredit, de faite, le centre commercial du monde et qu'en conséquence le nombre des dépêches extra-européens qui y arrivent ou qui en partent dépasse de beaucoup toutes celles du continent européen. Les membres de la Conférence seront peut-être étonnés d'apprendre que, sans tenir compte de Londres, dont la correspondance à elle seule représente les deux tiers de la correspondance britannique, tout le trafic extra-européen du reste de la Grande Bretagne est encore plus du double du même trafic de la totalité des États de l'Europe continentale. Or chaque dépêche des villes de province de l'Angleterre, de l'Écosse et de l'Irlande, d'un côté, avec les Indes, de l'autre, coûte un franc de plus que la dépêche de Londres. En outre, les autres centres principaux du commerce de l'Europe avec l'Asie orientale sont situés dans la partie occidentale du Continent européen», ibid., 471.

38 «M. Lasard ajoute que le trafic de l'Angleterre avec les Indes formant la presque totalité de la correspondance, l'établissement des lignes terrestres ou sous-marines indo-européens n'a pas été fait pour l'Europe continentale, mais pour l'Angleterre ; on ne saurait donc demander aux États de l'Europe qu'ils subissent des pertes pour favoriser cette correspondance », ibid., 474.

39 «Il veut répondre aux reproches adressés aux Administrations européennes de n'avoir rien accordé aux entreprises extra-européennes. Il reconnaît qu'à la vérité à Vienne quelques Offices leur ont demandé et ont obtenu d'elles des taxes trop réduites, mais, depuis ce moment, il rappelle que l'on a consenti aux augmentations que ces Offices réclamaient et qu'à St. Petersburg on a accepté d'emblée leur proposition d'établir le tarif par mot et d'adopter une taxe unique, ce qui est de la part des Administrations de l'Europe une sérieuse concession.», ibid., 475.

2.3 The private companies

A further sign of the gradual shifting of political weight within the TU was to be seen in the increasing numbers of private companies present, always reigned in by the TU, which kept true to its original policy and insisted on only admitting sovereign states as full members[40]. Yet the ever-growing role of the submarine cable companies in the construction of a global telegraph network meant that they could not be safely ignored[41].

At the Paris Conference, the Italian delegate Minotto proposed introducing an amendment to the article on the norms of adhesion to the TU (no. 59) in order to extend the right to private companies[42]. Followed by others, Vinchent felt it was formally unacceptable and spoke against it[43]. Minotto also proposed imposing surcharges on the companies not complying with the Convention's international rules[44]. In this case too,

---

[40] The 1858 Berne Conference allowed private companies to join the convention with the proviso they were represented by a sovereign state during proceedings «Les compagnies télégraphiques seront également admises sur leur demande à adhérer à la Convention. Dans ce cas elles ce poussant ce faire représenter à la Conférence par les délégués de l'un des États Contractants à leur choix», art. 1, *Convention Télégraphique de Berne 1858.*

[41] For a parallel study on the role of private companies within the ITU, see: Patricia K. McCormick, "Private sector influence in the International Telecommunication Union," *The Journal of Policy, Regulation and Strategy for Telecommunications, Information and Media* 9, no 4 (2007): 70-80.

[42] «Sur ce même article 59, M. le Délégué de l'Italie propose un nouvel amendement. Il demande qu'on substitue au premier et au troisième paragraphe de l'article 59 les deux paragraphes suivants : "Les compagnies concessionnaires de lignes télégraphiques seront admises à adhérer à la présente Convention. Cette adhésion emportera, de plein droit, accession à toutes les clauses et admission à tous les avantages stipulés par la présente Convention ; mais le tarif de l'article 30 ne s'appliquera aux correspondances des adhérents que s'ils adoptent des taxes proportionnelles sur leurs lignes», *Documents de la Conférence Télégraphique Internationale de St Petersburg*, 218.

[43] «M. Vinchent répond à M. le Délégué de l'Italie qu'en ce qui touche puisse être admise à adhérer à une convention internationale. Il ne voit pas d'ailleurs la difficulté dont M. le Chevalier de Minotto se préoccupe, les différents États s'étant en général réservé le droit d'imposer à leurs concessionnaires les dispositions réglementaires du service international», ibid., 219.

[44] «M. le Délégué de l'Italie avait proposé d'établir d'avance les conditions qui seraient faites par les États signataires de la Convention aux États non adhérents et aux compagnies privées qui refuseraient d'en accepter les clauses : les taxes fixées par la Convention ne seraient pas applicables à la correspondance de ces États et de

Vinchent and the majority of the delegates did not agree and held it was inappropriate to use any kind of force against administrations which did not recognize the convention[45].

So Paris confirmed that TU was still to be seen as an institution reserved for sovereign states, which could undertake to extend convention rules to the private companies they had signed contracts with. It was up to the single states to come to agreements with the private companies over adjusting tariffs to the charges established by the conference[46]. However, the companies refusing to recognize convention rules would not be able to share the tariff reductions agreed on at Paris[47]. So though the TU did not adopt any form of open coercion, it favoured the private companies which pledged to respect the Convention.

At the Vienna Conference the Italian delegation maintained its special relationship with the submarine companies and acted as spokesman for the Malta-Alexandria Cable Company, making public its desire to comply with the rules and consequently lower its charges to the Convention's agreed levels. Its adhesion could only be indirect, because Italy did not act as guarantor as called for by the Convention[48]. The episode well illustrates the difficulties encountered by the privates in adhering to the TU. Indeed, during the Vienna Conference, the question of the private companies was deemed of minor importance and the article regulating their adhesion to the convention was left practically unchanged. Significant, however, the

ces compagnies, et la Convention déterminerait la quotité des surtaxe dont leur correspondance serait frappée», ibid., 218.

[45] «M. Vinchent, répondant par avance à la proposition corrélative que M. de Minotto a reportée sur l'article 60, se déclare d'ailleurs formellement opposé à l'insertion de toute mesure coercitive à l'égard des compagnies. Il signale les graves embarras que l'on pourrait se créer dans l'avenir, en présence de compagnies qui seraient dans l'impossibilité réelle d'abaisser leurs tarifs», ibid., 219.

[46] «Les Hautes Parties contractantes s'engagent à imposer, autant que possible, les règles de la présente Convention aux compagnies concessionnaires de lignes télégraphiques terrestres ou sous-marines, et à négocier avec les compagnies existantes une réduction réciproque des tarifs, s'il y a lieu.», art. 60, *Convention télégraphique internationale de St. Petersburg 1875*.

[47] «Ne seront compris, en aucun cas, dans le tarif international : 1° Les bureaux télégraphiques des États et des compagnies privées qui n'auront point accepté les dispositions réglementaires uniformes et obligatoires de la présent Convention ; 2° Les bureaux télégraphiques des compagnies de chemins de fer ou autres exploitations privées, situés sur le territoire continental des États contractants ou adhérents, et pour lesquels il ya aurait une taxe supplémentaire», ibid.

[48] *Documents de la Conférence Télégraphique Internationale de Vienne*, 443-445.

fact that the conference unanimously approved a declaration proposed by the French delegate Jagerschmidt which called on the submarine cable companies to conform as far as possible to TU rules and tariffs[49].

Beginning with the Rome Conference, however, the question of private company membership became more important, given the presence of Great Britain and the consolidation of transcontinental cable companies[50]. Back in 1869, the secretary of the Bureau had insisted vehemently with the head of Italian telegraphs and organizer of the Rome Conference on the need to invite the private companies, especially those that were or would be supplying transatlantic telegram services. Once Rome was under way, the Italian administration presented a regulations project for the participation of private concerns[51], a topic that took up most of the first sitting[52]. Tornielli, an Italian Foreign Office official and chairman of the sitting, announced that he had invited two private companies[53], while leaving the Conference to decide if and how to allow them into the debates. Two questions emerged, the first concerning procedure and the second the substance. The norm admitting privates into the sittings had to be inserted in the Conference rules or directly in the Convention? If it had to go in the Conference rules, it would have to be re-voted at every conference, if in the Convention, an article would have to be modified and would only be definitely approved at the end of the Conference. And that

---

49 «La Conférence, amenée par le cours de ses travaux à s'occuper des compagnies de télégraphie privée, exprime l'admiration que lui cause l'énergique persévérance avec laquelle les promoteurs de l'entreprise du câble transatlantique ont poursuivi, au milieu de difficultés de toute nature, la jonction des deux mondes et réalisé cette grande œuvre de paix et de civilisation. Mais elle regrette que les règles adoptées par la compagnie pour l'exploitation de sa ligne s'écartent d'une manière si notable des règles usitées en Europe. Elle insiste spécialement sur les complications du tarif et émet le vœu qu'en ce qui concerne l'établissement des taxes, la compagnie se conforme, autant et aussitôt qu'il lui sera possible, aux dispositions de la Convention», ibid., 410.

50 "Conférence télégraphique de Rome (2ème article)," *Journal Télégraphique* 1 (1872) : 10-12.

51 «Les fonctionnaires attachés aux membres délégués et, le cas échéant, les représentants des Compagnies télégraphiques ou des experts peuvent être admis aux séances, après y avoir été autorisés par le Président pour chaque cas spécial, mais sans avoir le droit de prendre part à la votation», art. 4 du Règlement des Conférences de Rome, *Documents de la Conférence Télégraphique Internationale de Rome*, 236.

52 ibid., 228-236.

53 *Mediterranean Extension* and the *Anglo-Mediterranean Telegraph Company.*

meant the companies would have been excluded from Rome[54]. A compromise decision was taken, to modify the rule and evaluate at a later date whether to change the convention too and make a temporary choice permanent[55].

The substantial question was debated between two extreme positions. The Dutch delegate Staring wanted to allow the companies into the conferences with a consultative vote[56], while the Russian Luders, wanted to allow them only on the committees[57]. The German Meydam tried to mediate and argued that the company delegates could be allowed into the sittings they were most interested in[58]. Vinchent's proposal was another compromise: invite the companies to the conference providing they spoke via a common body[59]. In the end it was decided to allow the companies

---

54 «M. le Président croit qu'il se présente ici deux questions, d'abord, celle de l'admission des Compagnies aux Conférences de Rome, puis celle du principe de leur admissions à toutes les Conférences en général. Si l'on ne prend de décision que pour la réunion de Rome, c'est alors le règlement des Conférences qui doit enregistrer la solution. Veut-on, au contraire réant résoudre la question définitivement, il y aurait lieu de modifier la Convention», ibid., 229.

55 «M. Brunner et M. Staring ne pensent pas qu'il y ait nécessité d'attendre la discussion de l'article 66. La question peut dès maintenant être résolue en principe, quitte à modifier, au moment voulu, la rédaction des articles y relatifs de la Conventions, suivant la détermination adoptée», ibid., 230.

56 «M. Staring précisant alors sa pensée, propose de décider que les Compagnies sont admises aux Conférences avec voix consultative», ibid.

57 «M. Lüders serait d'avis, au contraire, de ne pas admettre les Compagnies dans les Conférences mêmes, mais seulement dans les Commissions, par l'intermédiaire desquelles elles pourraient faire parvenir aux Conférences l'expression de leur désirs», ibid.

58 «M. Meydam pense que la Conférence a plus de liberté, puis qu'elle peut, suivant les circonstances, autoriser, ou non, la présence des représentants des Compagnies. Quant au mode pratique de procéder, il serait bien simple. Les compagnies s'adresseraient au délégué de l'État dont elles relèvent qui à son tour, demanderait pour elles au Président l'autorisation d'assister à telle ou telle discussion», ibid.

59 «Il [M. Vinchent] propose à la Conférence de prendre une décision qui pourrait être formulée de la manière suivante : «La Conférence télégraphique de Rome, considérant qu'un certain nombre de représentants des principales Compagnies télégraphiques privées se trouvent en ce moment à Rome, ayant, d'ailleurs, égard à l'intérêt que les délibérations pourraient emprunter aux renseignements fournis par ces représentants. Considérant, toutefois que leur admission dans le local des séances ne doit, ni créer un précédent, ni prolonger indéfiniment les discussions par l'intervention d'un trop grand nombre de voix consultatives, prie le Président d'inviter à assister personnellement aux séances de la Conférence et des

access to all the debates. The chairman of the sitting had the right to admit all companies who made the request, but he passed this right over to the conference itself[60]. On the third sitting of 6 December 1871, the representatives of eleven private companies, most specializing in submarine cables, were present for the first time at a telegraph conference[61] .

Their presence made sparks fly between the two contrasting visions of the international telegraph service and the TU's role in regulating it. Many delegates took up a very conservative stand and spoke of incompatibility between the presence of private companies and the interests defended by the TU. During the discussion over the order in which to transmit telegrams, for example, the French delegate Ailhaud commented that the

---

Commissions les représentants des Compagnies qui ont été conviées à se faire représenter à Rome, en leur demandant de désigner parmi eux un organe unique chargé, sans voix délibérative, d'exprimer leurs vœux, de fournir à la Conférence toutes les informations utiles aux relations télégraphiques communes et de discuter, soit en son propre nome soit au nom de ses collègues, les propositions soumises à la Conférence», ibid., 231.

[60] «M. le Président fait observer qu'il résulte de ces différents votes que les représentants des Compagnies pourront être admis dans certaines séances et qu'une fois admis, ils auront tous la liberté de prendre la parole, mais sans participer aux votations. Il déclare de nouveau, d'ailleurs, qu'il ne compte pas user du pouvoir discrétionnaire que lui confère la rédaction adoptée et que son intention est de consulter toujours la Conférence sur l'opportunité des admissions, dans tel ou tel cas, des représentants des Compagnies. Il demande notamment s'ils doivent ou non être invités pour la prochaine séance», ibid., 235.

[61] «Sont présents, en outre, MM. Les représentants des Compagnies de lignes télégraphiques privées à savoir : Pour les sept Compagnies sous-marines unies : Falmouth, Gibraltar and Malta Telegraph, Marseilles, Algiers and Malta Telegraph, Anglo-Mediterranean Telegraph, British Indian Submarine Telegraph, British Indian Extension Telegraph, British Australian Telegraph, et China Submarine Telegraph, M. Jules Despecher, représentant également la Compagnie Construction and Maintenance Telegraph, Sir James Anderson, Directeur de la Compagnie China Submarine Telegraph; Charles Bompas et Leis Wells, Chef [sic.] de la Comptabilité du trafic des Compagnies; Pour la Compagnie Indo-European Telegraph, MM. Le Docteur Werner Siemens et C. William Siemens, Directeurs et W. Andrews, secrétaire de la Compagnie; Pour les Compagnies Great Northern Telegraph et Great Northern China and Japan Extension Telegraph, M. Edouard Suenson, lieutenant de marine; Pour la société du Câble transatlantique français et pour la Compagnie Anglo-americain Telegraph, M. Louis Curchod, administrateur délégué de la Société du Câble transatlantique français», ibid, 261-262.

interests of private companies and states were irreconcilable[62]. And at the end of a heated discussion over article 66 (then 64) about the adhesion of private companies, the Austrian representative Brunner observed that his initial evaluation had proved correct, that whatever decision was taken, the states would find themselves in total disagreement with private companies[63]. Finally the British India delegate Robinson labelled as disastrous the approval of some amendments to article 66 which gave greater guarantees to the submarine cable companies in comparison to overland ones, and he even reached the point of threatening to leave the TU[64].

Neither was there much that was conciliatory in the private companies, right from the beginning. For example, Despecher, the representative of five cable companies, intervened in the discussion over the order of telegraph transmission and highlighted not only an incompatibility of rules between cable and landline telegraphs but also accused the TU of imposing its rules[65]. He then expressed his disapproval over the scarce space given to the companies in the Conference and highlighted yet again

---

[62] «M. Ailhaud trouve que l'importance et la prolongation de la discussion prouvent la difficulté d'accorder entre eux les intérêts différents des Administrations d'États et des sociétés privées. […] Il y a là des divergences inconciliables, parce qu'elles partent d'un point de vue opposé», ibid., 287.

[63] «[M. Brunner] constate que la Conférence donne ainsi, plus tôt qu'il ne s'y attendait, une confirmation unanime à ce qu'il avait prévu, à savoir que, quelque décision que l'on prît, l'on se trouverait toujours, à un moment donné, en désaccord forcé avec les Compagnies», ibid., 402.

[64] «M. Robinson considère l'adoption de l'amendement de l'Italie relatif à l'article 66 comme une résolution funeste pour la Convention. […] Ceux qui ont voté pour cet amendement croient que chaque État a la faculté d'accorder des concessions aux conditions qui lui conviennent le mieux. Ceux qui ont voté contre admettent que les États qui avaient auparavant le devoir d'imposer les règles de la Convention, sont désormais déliés de cette obligation. Quant aux Compagnies, elles se considèrent comme libérées de tous leurs engagements et ne pensent même pas avoir besoin, à l'avenir, de fournir les documents justificatifs de leurs tarifs. Le résultat pour les correspondances avec les Indes sera d'avoir deux lignes terrestres liées par des obligations, en présence d'une ligne sous-marine indépendante. […] Si donc l'amendement, avec toutes les incertitudes qu'il soulève, devait être définitivement adopté, M. Robinson ne pourrait proposer à son Gouvernement de ratifier la Convention», ibid., 510.

[65] «Il [M. Despecher] ajoutera [...] que la Convention est imposée aux Compagnies sous-marines par les Gouvernements concessionnaires. Du moment que celles-ci ne sont pas libres de déterminer elles-mêmes les règles de leur exploitation, il lui paraît de toute justice qu'on leur accorde les réserves qui sont nécessités par les exigences spéciales de leur service», ibid., 288.

the financial and technological features of submarine telegraphs, which were ill adapted to some of the important but rigid TU norms[66].

Faced with these two extreme and apparently irreconcilable positions, some of the delegates tried to mediate in order to satisfy both state administrations and private companies. Curchod positioned himself midway between the two groups, enjoying both his dual status as representative of the *Société du Câble transatlantique français* and caretaker delegate for Switzerland, in the place of Lendi[67]. Curchod expressed the desire to create international regulations which could take into account both the particular features of the private companies, especially those dealing with cables, and the needs of the state administrations[68].

He felt that they were complementary within an international network in that they carried out functions which the other either could not, or

[66] «M. Despecher fait remarquer que le but que se proposent les Compagnies est très-clair ; elles désirent être maîtresses de leurs tarifs. [...]Si les Compagnies sont représentées ici, ce n'est pas pour mêler leurs débats à ceux des délégués des États, mais pour s'inspirer de la pensée des Conférences. [...]Il ne faut pas oublier que les Compagnies sous-marines sont des entreprises particulières qui nécessitent d'énormes capitaux. Elles ont le devoir de prendre les intérêts de leurs actionnaires et, dès lors, elles doivent rester seules juges des conditions de leur tarifs. Leur imposer des taxes immuables pendant trois ans serait rendre leur exploitation impossible, car leur existence est trop récente pour qu'elles sachent le régime qui convient le mieux, c'est-à-dire celui, en demandant au public la plus petite taxe, donne néammoins [sic], par la multiplication des dépêches, le produit le plus rémunérateur.», ibid., 394-396.

[67] Lendi was the caretaker secretary of the Bureau after Curchod's resignation in 1869. Following his death a few months later, Curchod took up his old job again.

[68] «Il espérait que l'on écarterait les différences qui existent entre les deux régimes ou que l'on prévoirait dans la Convention les dérogations nécessaires, pour obtenir par là l'accession des Compagnies américaines dans les conditions de l'article 66 et arriver, de la sorte, à n'avoir plus, pour toute la télégraphie, qu'une seule règlementation. En dehors, d'ailleurs, de toute préoccupation des intérêts des Compagnies et comme délégué de la Suisse, il lui semble que du moment où l'on offre au public la possibilité de diminuer le prix du service que lui rend la télégraphie à grande distance, il doit accepter avec empressement cette réduction de taxes élevées ; car, si les complications qu'elle peut introduire dans le service des Administration ne sont pas admissibles pour des correspondances très-nombreuses et à bas prix, l'inconvénient qui en résulte, lorsqu'il s'agit de dépêches beaucoup plus rares, ne saurait balancer les avantages que le public en retire», ibid., 371.

would not, touch[69]. A second intermediary position, nearer to the private companies, was held by the Italians, as always in favour of them being involved in the TU. The Italians felt that given the technological and financial differences, the cable companies had a right to be flexible and accept only the articles of the convention which were not explicitly against their own activity[70]. Hostile to this flexibility and a steadfast supporter of the need to sign the whole Convention[71], Vinchent held a third intermediary position, this time closer to the state administrations in spirit. Following on from the custom in force from the early period of the TU, he suggested the companies should accept or reject *in toto* the convention as it stood[72]. However, especially with regard to more delicate norms like

---

[69] «Au nome des Compagnies transatlantiques, M. Curchod doit déclarer qu'il ne leur serait pas possible d'accepter l'article 34, si l'on n'admettait pas en leur faveur quelques dérogations. […] Il faut, en effet, un grand trafic pour alimenter et faire vivre ces exploitations sous-marines, et, pour arriver au point qui donne le maximum de rendement, l'on doit passer par bien des tâtonnements. L'on a reproché aux Compagnies, ou pour parler plus exactement, l'on a invoqué contre elles la préoccupation qu'elles ont des intérêts de leurs actionnaires. Les conséquences que l'on tire de cet argument ne sont pas fondées. Les Compagnies accomplissent un but d'intérêt public. Elles se sont constituées pour combler les lacunes que les Gouvernements ne voulaient pas remplir; mais l'expérience ne leur a pas appris encore quel est le véritable tarif qui convient à la fois à leurs intérêts et à ceux du public. […] Si l'on croit désirable d'amener l'accession des Compagnies, il est donc nécessaire d'introduire dans l'article 34 une disposition qui leur laisse la faculté de faire des modifications de ce genre», ibid., 399-400.

[70] «Il [M. D'Amico] s'agit de savoir s'il n'est pas préférable d'avoir un grand nombre d'adhésions conditionnelles qui, pour la plupart des dispositions, équivalent, en fait, à des adhésions officielles, qu'un petit nombre d'accessions complètes», ibid., 369.

[71] «M. Vinchent ne reconnait pas cette distinction des adhésions conditionnelles qui n'est pas prévue dans la Convention. Il n'existe que l'accession pleine et entière sans réserve, que règle l'article 66, et un *modus vivendi* qui n'oblige que les États et laisse toute liberté aux Compagnies, celui qui fait l'objet de l'article 67.», ibid., 370; «Les compagnies sont libres d'accepter les règles de la Convention. A celles que leur acte de concession engage, il répondra qu'elles n'ignoraient pas, en souscrivant à ce contrat, les obligations qu'elles assumaient et qu'elles avaient, sans doute, des raisons sérieuses pour les accepter. Dans tous les cas, cette question ne regarde pas la Conférence qui n'a à transiger qu'avec des faits et non avec des désirs. C'est un fait qu'il est de l'intérêt du public d'obtenir des facilités qui diminuent pour lui le prix des correspondances lointaines. Mais, si l'on cédait sur des points où l'intérêt général est opposé à celui des Compagnies, la Conférence compromettrait l'œuvre qu'elle a mission de remplir», ibid., 397.

[72] «D'un autre côté, l'on objecte que les Compagnies se considèrent comme libérées de tout engagement. Cela dépend de leur acte de concession que le vote de

tariffs or adhesions, he suggested amendments which also took into account the needs of privates[73].

The contrasts between these different interpretations of the service went on right through the Rome Conference and surfaced particularly in the debates over the order of telegram transmission, tariffs and private company adhesions. At the end of the debate on article 66, Despecher expressed his appreciation of the positions of both the Italians and Vinchent, but required some guarantees of tariff flexibility for private companies[74]. Vinchent's line was the clear winner. More traditional and closer to the state administrations, it was shared by the majority of delegates. Yet in spite of the many compromises made to encourage the private companies to enter, they chose not to:

> The delegates of the telegraph companies united for the service of India, China and Australia regret to have to inform you that after due consideration they find it their duty not to make a closer act of adhesion to

---

la Conférence n'a pu annuler. Elles croient également avoir tout intérêt à conserver leur liberté. Il y a cependant des avantages sérieux en faveur des Compagnies adhérentes. Ces dernières ont le droit que les tarifs, une fois arrêtés sur la ligne qu'elles desservent, ne puissent plus être modifiés sans leur participation; elles ont le droit d'être informées de ces modifications dans les délais voulus; elles ont droit au bénéfice de la transmission gratuite de la mention de la voie. Aux Compagnies indépendantes, au contraire, les États ne doivent rien et peuvent ne leur accorder que ce qui leur convient, pour sauvegarder les intérêts de leur public», ibid., 513 ; «En résumé, les Gouvernements ont, comme auparavant, le droit et le devoir d'agir dans le sens de l'acceptation de la Convention, et si les concessions qu'ils ont données les laissent désarmés à cet égard, il reste encore à leur disposition toutes les mesures de faveur qui rendent préférable pour les Compagnies le régime de l'accession à celui de l'indépendance. Si cette opinion n'était pas celle de la majorité des membres de la Conférence, M. Vinchent prierait ceux d'entre eux qui ne la partagent pas de se prononcer à cet égard», ibid., 514.

[73] See the amendments proposed on art. 66 by Vinchent, ibid., 597-598. «Bien que ces diverses appréhensions ne lui paraissent pas justifiées, M. Vinchent ne voit aucun inconvénient, pour calmer les scrupules qui se sont produits, à admettre le troisième paragraphe qu'il a proposé, paragraphe dont, pour lui-même, il ne sent pas la nécessité», ibid., 604.

[74] «Dans leurs rapports avec les États contractants, les Compagnies se trouvent en présence de deux opinions différentes : les uns, comme M. le délégué de la Belgique, croient préférable d'arriver à une entente plutôt que d'accorder à chaque Compagnie sa liberté d'action. Les autres, comme M. le Président, aiment mieux laisser les Compagnies se mouvoir librement. Ce sont deux modes d'agir différentes, mais qui seraient acceptables l'un et l'autre, pourvu que l'on tînt compte des nécessités des Compagnies», ibid., 496.

the Convention than in the past. They consider that the High Contracting Powers' desire for uniformity infringes the rights of the private companies without any equivalent compensation being offered, in return for their loss of commercial freedom that the Governments of their countries will always have close to their hearts to preserve.

It was already because of these impediments to exercising the same principle that the Companies had refrained from signing the Vienna Convention. But having been invited to the Rome Conference they have come with the idea that new changes in their favour would be introduced to facilitate their membership. This hope has not been entirely realised. They feel that on the contrary the tendency of several amendments has been to make more restrictions to the Companies' freedom of action than at Vienna and that when modifications have been proposed, the wishes of the Companies have been subordinated to protecting opposing interests and increasing government regulations on the commercial lines which have contributed so much to prosperity. We think that the Convention has not understood the exceptions that separate us and that the clause they are applying leaves our shareholders with no privilege except that of having spent more that a hundred million to install submarine cables, whose revenue will be regulated and hindered by dispositions established without their taking part in the voting, and in certain cases without any considerations except uniformity of government convenience[75].

---

[75] «Les Délégués des Compagnies de télégraphie unies pour le service de l'Inde, de la Chine et de l'Australie, regrettent d'avoir à vous faire part qu'après sérieuse considération, ils jugent de leur devoir de ne pas faire acte d'adhésion plus intime que par le passé à la Convention. Ils considèrent que le désir de l'uniformité de la part des Hautes Puissances contractantes l'a emporté sur les droits des entreprises privées, sans qu'il ne leur ait été concédé aucune compensation équivalente, en retour de l'abandon de la liberté commerciale que le Gouvernement de leur pays aura à cœur de leur conserver. C'était déjà en raison des entraves à l'exercice de ce même principe que les Compagnies s'étaient abstenue d'adhérer à la Convention de Vienne; mais ayant été invitées à la Conférence de Rome, elles y sont venues avec l'idée que de nouvelles modifications, en faveur de leur service, seraient introduites en vue de faciliter leur accession. Cet espoir ne s'est pas complètement réalisé. Elles considèrent, au contraire, que la tendance de plusieurs amendements adoptés dans la révision actuelle a été d'apporter plus de restrictions à la liberté d'action des Compagnies que les dispositions de la Convention de Vienne, et que, lorsque des modifications ont été proposées, les vœux des Compagnies ont été subordonnés à la protection d'intérêts opposés et à l'accroissement de la règlementation gouvernementale sur les lignes commerciales qui ont tant contribué à la prospérité des nations. Nous pensons que la Convention n'a pas bien compris les quelques exceptions qui nous séparent et que les clauses auxquelles elles s'appliquent ne laisseraient à nos actionnaires d'autre privilège que celui d'avoir pensé plus de cent millions pour l'établissement de câbles sous-marins, dont les revenus seraient régis et entravés par des dispositions établies sans leur

After working hard to push through a series of amendments in favour of the private companies, Vinchent was understandably left speechless[76]. The Rome Conference thus ended on an extremely ambivalent note. Article 64 indeed openly favoured the submarine cable companies over the private land ones, which were obliged to accept the Convention if they operated within a member state. And yet the submarine cable companies had manifested their desire not to be subject to the Convention's rules. A prelude, indeed, to a new balance of power between state administrations and private companies.

During the four years that lay between Rome and St Petersburg, the submarine cable companies increased considerably in importance and industrial concentration, thanks above all to the British businessman John Pender, who took over and merged many companies[77]. Acquiring and preserving a monopoly over a given area therefore became an element of sharing and cohesion between public administrations and submarine companies.

In fact, the battle-line between private and government interests in Rome almost seemed to fade away at St Petersburg. There was only one communication against private companies, but in this case too, the German delegate Stephan pointed to defending monopolies both for governments and private cable companies. Furthermore, when Despecher, now delegate for the Pender company Eastern Telegraph, declared his desire to join the Convention and asked it to extend to private companies the anti-dumping norm which for many years had protected state administration from competition[78]. His proposal was seconded by James Anderson[79], a highly

---

participation au vote, et, dans certains cas, sans autre considération que celle de l'uniformité ou celle de la convenance des Gouvernements», ibid., 609-610.

[76] «M. Vinchent ne s'explique pas la raison d'être de cette déclaration. En ce qui le concerne, il ne la comprend pas et il ne voudrait pas que son acceptation fût considérée comme impliquant le fait qu'elle a été comprise par la Conférence», ibid., 612.

[77] Peter J. Hugill, *Global Communications since 1844. Geopolitics and Technology* (Baltimore: Johns Hopkins University Press, 1999), 25-52.

[78] «Par un heureux emprunt des expressions de la littérature française classique, l'on a dernièrement ici comparé avec beaucoup d'esprit les Compagnies des câbles à des belles. M. Stephan s'associe de grand cœur à cette comparaison, qui lui permet de se prévaloir d'un fait, que personne ne voudra contester : c'est que le cœur de l'homme a toujours un faible pour les belles et il prend la liberté de rappeler au souvenir de l'assemblée que dans la présente Conférence, les représentants des États ont bien montré qu'ils ne faisaient point exception à cette loi universelle à la quelle tout le sexe masculin est soumis. Mais, une des qualités principales de la beauté, un de ses attraits mystérieux, réside dans la rareté.

esteemed expert in the economy of communications. Anderson held that reducing international tariffs was detrimental in that it would not produce the corresponding increase in traffic that had happened with the domestic service[80]. Vinchent backed Despecher by holding that the anti-dumping law was extended automatically to all TU members, so that signing the St Petersburg Convention meant automatically accepting it[81]. Despecher then

---

Maintenant, M. Stephan le demande, est-ce que les Compagnies n'ont pas un peu perdu de cet attrait de la rareté ? Et cependant, on en veut toujours davantage! Les chercheurs de concessions de nouveaux câbles poursuivent les Gouvernements; c'est ce que tous les Directeurs généraux des télégraphes réunis dans cette enceinte reconnaîtront unanimement, à la seule exception de leur collègue de la Suisse, puisque le territoire helvétique ne touche pas à la mer; ce qui est, dans le cas présent, un véritable avantage. [...]Si des conjonctions tellement accidentelles et fragiles peuvent amener de tels résultats, il faut bien avouer qu'un système qui les provoque ne correspond plus aux rapports existants à notre époque. Car voici ce qui peut arriver encore. Il se produit une rupture du câble, et tout-à-coup la masse entière des télégrammes se précipite sur les lignes du pays qui offre l'intermédiaire naturel et dont la voie est réellement la plus courte. L'Administration de ce pays, à qui ce travail avait été tout-à-fait enlevé par la voie détournée, n'a plus le nombre suffisant de fils et d'employés à sa disposition, et pourtant elle doit faire face à cet encombrement subit qui se présente résulte évidemment des perturbations dans son service régulier. Trouve-t-on juste de détourner le transit entier d'un pays qui constitue la route directe? Trouve-t-on équitable qu'il doive se contenter du rôle indigne de pis-aller qu'on lui impose? M. Stephan pense que le raison et plus encore le sentiment des membres de cette assemblée ne leur permettront pas de répondre à ces questions autrement que par un non», *Documents de la Conférence télégraphique internationale de St. Petersburg*, 495-497.

79 «L'expérience a établi en télégraphie deux axiomes qu'on ne devrait jamais perdre de vue, surtout dans les questions de concurrence entre exploitations privées. Le premier, c'est que toute réduction de tarif a pour résultat immédiat de réduire les recettes. Le second, généralement moins bien compris, c'est que les réductions n'ont pas sur le développement de la correspondance internationale ou s'il s'agit de la correspondance à l'intérieur d'un État ou d'une grande ville. Cette différence d'analogie existe à un plus haut degré encore pour la correspondance extra-européenne. Aucune augmentation très notable dans le nombre des dépêches ne peut résulter d'une réduction, d'autant moins que pour cette nature de correspondance la réduction ne peut jamais être que peu considérable relativement au total», ibid., 556.

80 Anderson had already given an account of his convictions in: James Anderson, *Statistics of Telegraphy. Read before the statistical society June 18th 1872* (London: Waterlow and Sons, 1872).

81 «M. Vinchent explique qu'en se prononçant pour l'accession à la Convention. Il va sans dire que l'on comprend par là l'adoption complète de ses règles et de ses principes, en matière de tarification comme ailleurs. La disposition relative á la

confirmed Eastern Telegraph's adherence, and implicitly that of the submarine cable companies, to the "great telegraph family"[82]. Using the anti-dumping norm to defend the monopoly turned out to be sufficient to unify the aims of private companies and public administrations[83]. This unity of purpose may well be identified as a basic characteristic of the TU, which brought it to back an economic-managerial *status quo* in the telecommunications market[84].

2.4 The identity of the delegates

Following the tradition of the previous conferences, in Paris 1865 states were represented by mixed delegations composed of at least one diplomat plus a technician, normally the head of national telegraphs. From London 1879, which followed St Petersburg, all conferences were administrative, and therefore composed exclusively of delegations from the various administrations, often the civil servants running the services.

In theory, the conferences between Paris and St Petersburg were all plenipotentiary, though in practice in post-Paris conventions, technical experts debated rules and tariffs by themselves. For example, in the first Paris sitting the French foreign minister Drouyn de Lhuys highlighted the fact that since the technical nature of the reforms called for experts, he was

---

non-concurrence ne vise, il est vrai, que les changements de taxes introduits hors des Conférences pour des voies existantes; mais l'esprit de la Convention est évidemment que les mêmes principes servent de guide pour la fixation des tarifs des voies nouvelles», *Documents de la Conférence télégraphique internationale de St. Petersburg*, 565.

[82] «En faisant acte d'adhésion à St. Pétersbourg, les Compagnies vont entrer dans la grande famille télégraphique. Elles sentiment, dès lors, qu'une protection égale doit s'étendre sur tous les membres de la famille, et garantir les Sociétés privées come les États», ibid.

[83] «Les modifications prévues au paragraphe 4 de l'article de la Convention devront avoir pour but et pour effet, non point de créer une concurrence de taxes entre les voies existantes, mais bien d'ouvrir aux public à taxes égales autant de voies que possible et les combinaisons nécessaires seront réglées de telle manière que les taxes terminales des Offices d'origine et de destination restent égales, quelle que soit la voie suivie», Art. XIV, *Règlement de service international, annexé à la Convention Télégraphique Internationale de St. Pétersbourg*.

[84] Peter Cowhey, "The International Telecommunications Regime: The Political Roots of Regimes for High Technology," *International Organization* 44, no 2 (1990): 169-199.

setting up a special committee[85]. What actually happened was that the plenipotentiary delegates, who were often professional diplomats, took part in the opening sitting and then only came back on 13 April to approve and sign the Convention drawn up in the 16 plenaries. During the closing proceedings, Drouyn de Lhuys underscored how the Convention was the fruit of the lengthy collaboration between telegraph officials[86]. For this reason already at the next Conference in Vienna many countries gave up sending double delegations and simply granted their heads of telegraphs[87] the power to sign the internationally binding treaties. By Rome[88] and St Petersburg[89] it had indeed become the custom. So the shift from diplomatic to administrative conferences was more a recognition of a state of affairs than the result of consciously chosen, carefully and cautiously debated norms.

In the terms of the times, both the diplomats and administrators taking an active part in the conferences represented the telegraph administrations of the member states. Yet the terms leave many questions hanging: 1)

---

[85] «Mais les nombreuses questions qui se rattachent à la réforme qu'il s'agit d'accomplir exigent un examen détaillé, ainsi que des connaissances techniques. C'est pour préparer la solution de ces questions que les divers Gouvernements ont bien voulu envoyer à Paris des Délégués spéciaux. M. Drouyn de Lhuys propose donc de former une Commission, composée de MM. les Délégués, qui serait chargée de soumettre à la Conférence un projet de Traité général.», *Documents Diplomatiques de la Conférence Télégraphique Internationale de Paris*, 77.

[86] «Ce ne sera pas sans avantages réciproques que des hommes d'élite, placés à la tête de grands service publics dans leurs pays, seront venus, de tous les points de l'Europe, mettre en commun les résultats de leurs expérience et constituer une sorte d'enseignement mutuel de haute administration», ibid., 306.

[87] «Aujourd'hui qu'il s'agit d'un travail de nature essentiellement technique, les Gouvernements, auxquels j'ai adressé des invitations conformément à l'article 56 de cette même convention, ont partagé ma manière de voir en déléguant de préférence des Représentants des Administrations respectives», *Documents de la Conférence Télégraphique Internationale de Vienne*, 92.

[88] «Des savants renommés, des hommes spéciaux ou d'une grande expérience administrative ont été appelés au secours des diplomates de profession ; et cette introduction dans la diplomatie d'hommes doués des plus précieuses connaissances positives, me paraît constituer une des innovations les plus fécondes dans les rapports des nations entre elles», *Documents de la Conférence Télégraphique Internationale de Rome*, 215.

[89] «La Conférence, composée de délégués administratif, préparera la Convention, qui devra être revêtue ensuite de la signature des plénipotentiaires de tous les États contractants. Cette Convention sera complétée par un règlement dont l'élaboration et la signature sont exclusivement dans les attributions des délégués administratif», *Documents de la Conférence Télégraphique Internationale de St. Petersburg*, 278.

Were the participants operating on behalf of their governments or were they biding by criteria of general efficiency? 2) Were they to consider themselves diplomats, technicians or simply managers?

A couple of significant episodes give food for thought about the role of the delegates. In Rome 1871, Field, one of the main promoters of the first transatlantic cable together with Morse, wrote to the delegates and invited them to come to an agreement to stop cables being cut in times of war. Vinchent interpreted the common feeling when he answered that the TU could only deal with administrative issues, not questions of international politics[90]. In the same way in St Petersburg, when the chair announced that some inventors desired to present their prototypes, they were told that the conference did not deal with technological innovations[91]. Evidently the delegates did not see themselves either as diplomats or technicians in a narrow sense, but more in the role of government officials.

At the St Petersburg Conference, there emerged a more precise definition of the conferences as "meetings of experts", where experts were perceived as: "intelligent men, ready to be lit up by debating and modify their opinion following the considerations that they hear being developed."[92]

They were in short telegraph managers and as such were meant to give their professional opinions. Though this idea was widely shared, it met

---

[90] «A cette occasion, M. Vinchent attire l'attention sérieuse de la Conférence sur la situation toute spéciale faite aux Administrations télégraphiques qui seules peut-être parmi tous les autres services publics, ont la faculté de traiter directement les questions internationales qui les intéressent le plus. Cette situation, il importe de ne pas la compromettre et, pour cela, il faut éviter de sortir du domaine purement administratif et de se lancer, sous forme de vœux ou autrement, dans des délibérations qui, par leur nature politique appartiennent à un autre ordre d'idées», *Documents de la Conférence Télégraphique Internationale de Rome*, 247.

[91] «M. le Président donne connaissance de deux communications qu'il a reçues, l'une d'un industriel belge recommandant un moteur destiné à remplacer les piles, et l'autre du sous-directeur à l'institut météorologique de Copenhague appelant l'attention de la Conférence sur un nouvel appareil rapide de son invention. Sur l'observation que la Conférence n'est pas une réunion de techniciens chargés de l'examen des inventions ou perfectionnements, le Bureau de la présidence est invité à répondre à ces communications, en faisant connaître qu'il ne peut leur être donné suite», *Documents de la Conférence Télégraphique Internationale de St Petersburg*, 561.

[92] «Hommes intelligents, prêts à s'éclairer par la discussion et à modifier leur opinion suivant les qu'ils entendent développer», ibid., 295.

some highly articulate opposition. For example, according to the Swiss delegate Hammer:

> The concept of the Conference as a meeting of experts is, furthermore [...] an incomplete concept, for there are the delegates and the first among them, who while not being experts on telegraphs have not fewer qualities for defending and upholding the interests of their countries with their words and votes[93].

These contrasting views of the conferences as meetings of experts on one side and government delegates on the other were aired and re-aired through the following sittings[94], right up to the point in which Hammer questioned even the concept of administrative conferences:

> M. D'Amico said they were simply administrative conferences. But in these conferences the delegates will not only have to deal with questions of service where experience is precious, but also the interests of their countries, as in the discussions over tariffs[95].

In reality, their role as experts meant representing their governments, too. All the delegates were *de facto* top officials of their public administrations and as such presumably great experts of the service they headed as well as loyal servants to their governments' briefs. The care to detail and precision with which delegates like Curchod, Vinchent,

---

[93]«La conception de la Conférence comme une réunion d'experts, est, d'ailleurs [...] une conception incomplète, car il y a des délégués, et lui-même tout le premier, qui, tout en n'étant pas des experts au point de vue télégraphique, n'en ont pas moins qualité pour défendre et soutenir les intérêts de leur pays par leur parole et par leurs votes», ibid., 296.

[94] «M. Brunner trouve que cette proposition ne répond pas à la conception des Conférences qui doivent être considérées comme des réunions d'experts. Il craint qu'une semblable innovation n'ait des conséquences fâcheuses et il voudrait connaître les motifs qui ont inspiré les auteurs de l'amendement. [M. Hammer] Quant à la conception de la Conférence comme une réunion d'experts, elle n'est point complètement juste, car les délégués n'apportent pas seulement le concours de leurs lumières, mais ils ont aussi la défense des intérêts qu'ils représente. Pour deux États, ces intérêts peuvent être d'une nature toute différente, tandis que pour deux Administrations d'un même État ils sont plus naturellement semblables», ibid., 511-512.

[95] «M. D'Amico dit qu'il s'agit simplement de Conférences administratives. Mais dans ces Conférences les délégués n'auront pas seulement à s'occuper de questions de service où les expériences sont précieuses, mais aussi, comme dans la discussion des tarifs, des intérêts de leur pays.», ibid., 513.

D'Amico and Brunner took the word in order to improve technical and managerial aspects of the international service reveals by itself their high level of competence. Nevertheless, it cannot be denied that they were very much creatures of their times and often moved in response to political briefs, which were always prioritized over technical matters. The tendency to defend one's government's interests was clear in all the debating, though there were some incidents which made it emerge more sharply. When in Vienna in 1868 Curchod was given the post of Bureau Secretary and then in St Petersburg managed to put in the hands of the Bureau (and therefore his) the entire conference process for approving norms, it was obvious that he was following his government's instructions. And those aimed at giving Switzerland a position of prominence by means of an adroit use of its much acclaimed neutrality[96]. In the same way the German delegate Meydam's spoiling for trouble and continuous communications on all matters were strategies designed to reflect Germany's power after its recent unification. Then there is the more discreet case of the Italian delegates, who between 1865 and 1875 followed their government's brief attentively and promoted the interests of the submarine cable societies whose international lines crossed the peninsula and brought so much important revenue into the Italian coffers.

The ambivalent role of the delegates must be taken into consideration in any analysis of the conventions, regulations and tariffs. With the most fundamental issues (tariffs, private company membership, anti-dumping norms), the delegates followed to the letter the guidelines laid out by their governments, while in working out more technical and marginal norms, they mostly relied on their own managerial experience. Whatever, there was indeed always reciprocal contamination between their two roles.

2.5 Competences and procedures

The conferences were necessarily periodic because the documents they produced were subject to constant revision, made necessary by the rapid expansion of the telegraph service and equally fast evolution of technology. It was indeed the unforeseeable speed of development of international telegraphs that lay behind the decision to establish the conferences at regular intervals. The delegates chose to meet whenever the technical or political conditions called for a revision of the convention and

---

[96] Simone Fari, Gabriele Balbi, Giuseppe Richeri, "The Bureaucratisation of the Telegraph Union," *Storia Economica* XVI, no 2 (2013): 377-394.

its rules and tariffs[97]. This is not actually what happened because, as will be later shown, a long term feature of the TU was to make do with its stagnating norms and technology. Therefore, in order to pick up all the nuances in the contents of the conventions, we need to analyze the procedure that lay at their base, at how the drafting and approval procedures underwent a gradual modification and from Paris up to St Petersburg were subject to a bureaucratization process quietly gnawing away at the power of the single delegates.

In all the conferences between 1865 and 1871, the convention agenda was drawn up by the host on the basis of general indications received from other administrations. At Paris the first draft of the convention and rules were discussed article by article with the relative amendments added by the delegates. At Vienna, though, Curchod proposed and obtained faster proceedings and accepted only amendments backed by at least two delegations[98]. Again in the name of streamlining, a committee was created to revise the regulations[99]. Three years later in Rome the committees had become three: a) tariffs; 2) regulations; 3) convention[100]. Another change in procedure was that amendments proposed by private companies were accepted for discussion.

The turning point, the takeover in the bureaucratization process, followed later, at St Petersburg. In line with the mandate received at the Rome Conference[101], Curchod's Bureau drew up the draft of the convention and rules.

Probably in order to tackle the needs concerning the formulation of a definitive convention, the St Petersburg Conference differed in both the preparatory stage and the assembly. During the first sitting, Curchod explained carefully what had happened in the phase of documentation preparation:

---

[97] «Il [M. Brunner] a pu constater que, soit après la première conférence de Paris, soit après celles de Bruxelles et de Berne, on n'a pas éprouvé la nécessité de se réunir aux époques que chacune de ces conférences avait fixées. Il croit par conséquent inutile de prévoir dès à présent une époque déterminée: son amendement a le double but d'éviter des réunions inopportunes, et de permettre celles que les circonstances pourraient rendre nécessaire, dans un délai même très-rapproché», *Documents Diplomatiques de la Conférence Télégraphique Internationale de Paris*, 212.

[98] *Documents de la Conférence Télégraphique Internationale de Vienne*, 215.

[99] Ibid., 261

[100] *Documents de la Conférence Télégraphique Internationale de Rome*, 227-228.

[101] Ibid.

M. Curchod […] announces that he has submitted first of all to the Russian Administration, as tasked by the present Conference organization a convention pre-project, which has been sent to all the states, as an attachment to the Imperial Government's circular.
This first project, in which the International Bureau in line with the Russian Administration's idea, only included the essential dispositions, consecrated in some way by a continuous experience and general adhesion, has undergone, following the codification of the regulations, some further modifications which do not distance from the principles adopted for its elaboration. The results of this double work were sent to all Administrations in January 1875, to be studied before the Conference. Mr Curchod notes about this that the examination will show that the Paris Convention dispositions, as coming out of the Rome deliberations, have been reproduced in the new convention and regulation projects. As for the propositions of the different Offices, they have only been produced in following the old text and the International Bureau had to gather and co-ordinate them in two different books, one of which, the yellow book follows the order of the subjects in the Rome text and presents them exactly as the Administrations themselves had produced them. The other blue book classes them in the order of the new projects, only adding formal modifications necessary in re-touching the original plan. Reference numbers create a concordance between the two books.[102]

---

[102] «M. Curchod [...] fait connaître qu'il a soumis, d'abord, à l'Administration russe, comme chargée de l'organisation de la Conférence actuelle, un avant-projet de Convention qui a été communiqué à tous les États, en annexe à la circulaire du Gouvernement impérial. Ce premier projet, où le Bureau international, d'accord avec les idées de l'Administration russe, n'avait compris que les dispositions essentielles, consacrées, en quelque sorte, par une expérience constante et une adhésion générale, a subi, à la suite de la codification du Règlement, quelques modifications ultérieures qui ne s'écartent point, d'ailleurs, des principes adoptés pour son élaboration. Le résultat de ce double travail a été transmis, au mois de janvier 1875, à toutes les Administrations, pour pouvoir être étudié avant la Conférence. M. Curchod fait observer à ce sujet que cet examen permet de constater que toutes les dispositions de la Convention de Paris, telle qu'elle est sortie des délibérations de Rome, ont été reproduites dans les nouveaux projets de Convention et de Règlement. Quant aux propositions des différent Offices, ceux-ci n'ont pu les produire qu'en suivant l'ancien texte et le Bureau international a dû, dès-lors, les réunir et les coordonner dans deux cahiers différents, dont l'un, le cahier jaune, suivant l'ordre des matières du texte de Rome, les présente exactement comme les Administrations elles-mêmes les avaient produites, et dont l'autre, le cahier bleu, les classe suivant l'ordre des nouveaux projets, en apportant seulement les modifications de formes nécessitées par le remaniement du plan primitif. Des chiffres de renvoi établissent la concordance entre l'un et l'autre de ces cahiers», ibid., 285.

From a rapid comparison between the acts of the St Petersburg Conference and the previous ones, it emerges that discussions over various convention and regulation articles were decidedly shorter. Most of the norms were read and approved without giving rise to the heated debates that had made sparks fly in the past. This because of a rigorous procedure conducted by the Bureau, which organized a series of preliminary phases designed to obtain the delegates' tacit agreement, and led to sittings where there was no need for debating or diverging. In the first sitting Curchod presented the project together with the amendments from the various delegations and suggested a procedure for approving the acts, which with some minor changes was accepted. The procedure provided for: 1) an initial reading of the project for the Convention, article by article; 2) a direct approval of the articles or re-reading of the regulation in the case of disagreement; 3) a reading of the regulation articles, one by one; 4) in the case of agreement, direct approval of the articles of the regulations and the convention norms pending from phase 2; 5) in the case of disagreement over an article, deferment to the relative committee[103]; 6) committees formulating proposals for new articles; 7) after the initial reading of the convention and relations, presentation of committee's opinion of articles pending on phase 5 (approval/rejection of changes); 8) in the case of the committee's proposal being judged favourably by the delegates, direct approval of the article; 9) in the case of the committee's proposal being judged negatively, opening of a discussion between the different positions and in the case of no agreement, article referred to Bureau; 10) collection of divergent opinions, the Bureau formulating a compromise; 11) almost automatic approval of the article as formulated by the Bureau.

All this complex bureaucratic procedure concentrated much power in the Bureau's hands. It had drawn up the convention/regulation project, collected the relative amendments and established the procedure for debating the articles, as well as acquired the power to close the approval procedure of the norms in the case of disagreement among the delegates. If an article was not approved during the sittings, phase 10 actually allowed

---

[103] The committees were bodies already formed in the Rome Conference to speed up discussion and approval proceedings of convention and regulation articles. While however in Rome they were informal with a prevalently consulting power, in St Petersburg their function was to formalize. Two committees were named during the first sitting, the first for tariff matters while the second focused on doubts about regulation articles. In St Petersburg the committees acted with the power of parliamentary ones and gave their opinion on articles which had aroused discord among the delegates. A committee's opinion was normally automatically accepted by the assembly.

the Bureau to formulate a compromise article keeping account of the needs of all.

2.6 Conclusions

The conferences were the determining component in the mechanism of the TU in that they gave birth to the norms and criteria for their application. In one way they were very innovative and in others extremely conservative and traditionalist. As we have seen, they were indeed a great political innovation, the first legislative body belonging to an international organization. However, while they clearly drew inspiration from legislative assemblies of a liberal cast, they also had to keep in mind new elements: 1) representatives were not elected but state-appointed; 2) delegates had to be both diplomats and technical experts; 3) matters under discussion would involve private companies too, given the international nature of the service.

Nevertheless, the conferences also inherited directly the tradition of the gestation period of the TU. First of all, both the multilateral treaties and the first telegraphic unions provided for a collegial body to draw up the convention and rules. Secondly, two of the 1848-1865 constitutive choices had a decisive influence on the nature of the conferences: 1) the domination of politics; 2) Eurocentrism. The fact that all the previous stages of telegraphic multilateralism had been under single sovereign states meant that the TU stood as an inter-governmental organization and closed off to private companies as effective members. In a similar way, though Great Britain's membership partially changed the internal balance, it did not affect the traditional continental powers and the perpetuation of a clear Eurocentrism. Finally, the progressive bureaucratization of the procedures for drawing up and approving the norms, which slowed down any introduction of innovations, only led to consolidating the TU's innate inertia.

The presence of this basic ambivalence in the conferences acting both as an element of conservation and at the same time rupture needs to be taken into consideration in any analysis of the numerous internal elements (diplomats/technical experts, private/public, administrations/states) if we want to acquire a depth understanding of how norms were drawn up and approved, and how tariffs were calculated and applied.

CHAPTER THREE

THE INTERNATIONAL BUREAU

The task of the Bureau was to maintain contacts between member administrations and carry out everyday business in the periods between conferences. The Bureau, unforeseen in the Paris Convention, was an absolute novelty and was set up only after the skilful operation carried out by the Swiss delegation during the 1868 Vienna Conference. Its origins are described in the first section below, while an analysis of its formal duties and real power of influence follows in the second section.

3.1 The Institution of the Bureau

The French Administration had originally inserted in the draft of the Paris Convention the establishment of a special international committee empowered to carry out administrative functions in the interval between conferences[1]. Some delegates were against the idea because they thought it would be an unnecessary outlay of resources and money for the Union[2]. To win more support, the French proposed holding committee meetings at

---

[1] «Une Commission composée des Délégués de chacune des Administrations télégraphiques des États contractants sera chargée de dresser la carte générale des relations télégraphiques, de publier les tarifs communs et de procéder aux études d'utilité commune dont elle pourrait être saisie. Cette Commission fonctionnera dans la capitale de l'État où la dernière conférence aura été tenue, et sous la direction du chef de l'Administration télégraphique de cet État. Elle sera chargée notamment de préparer les éléments de l'instruction mentionnée en l'article 54 ci-dessus, ainsi que les modifications qu'il pourrait devenir d'y apporter ultérieurement», art. 57 du Projet de Convention Internationale, in *Documents de la Conférence Télégraphique Internationale de Paris*, 83-107.

[2] For example, the Swedish delegate Brandstrom, felt that «le travaux d'utilité commune qui peuvent se placet entre deux Conférences ne sont pas assez important pour motiver la réunion permanente de délégués de chaque État». Furthermore, «M. Brunner s'associe à la pensée de M. Brändstrom et s'attache à montrer que les travaux que la Commission internationale aurait à effectuer ne motivent pas suffisamment la constitution de cette Commission», ibid., 216.

intervals and setting up a permanent office run by an employee of the host state. Yet again most of the administrations judged the creation of a permanent body inopportune, and the French idea was rejected[3].

Article 57 of the Paris Convention determined that the various administrations were to exchange notice of line interruptions, office openings and in general all variations in international communications[4]. During the three-year interval between the Paris and Vienna Conferences, the international telegraph network expanded so much that the system foreseen in the 1865 convention proved insufficient. At Vienna, therefore, two amendments were presented, one for the institution of a permanent body, the other for a semi-permanent structure. The latter, a French proposal, calqued the project presented at Paris of an all-member-state committee, which could meet in cases of doubts rising over how to apply the Convention[5]. Curchod, however, started from the assumption that the

---

[3] «M. le Président fait observer à le Délégué de la Suède que tel n'est pas le sens de l'article : la Commission proposée par le Projet serait en effet permanente, en ce sens qu'elle aurait à accomplir une mission continue ; mais cette mission pourrait être remplie sans que la présence de délégués de chaque État fût toujours nécessaire. Des réunions périodiques y suffiraient, et, pour ces réunions mêmes, les divers groupes d'États pourraient s'entendre et confier un mandat commun à un seul délégué», ibid.

[4] «Les Hautes Parties contractantes, afin d'assurer, par un échange de communications régulières, la bonne administration de leur service commun, s'engagent à se transmettre réciproquement tous les documents relatifs à leur administration intérieure et à se communiquer tout perfectionnement qu'elles viendraient à y introduire», art. 57, *Convention Internationale Télégraphique de Paris 1865.*

[5] «A cet effet il [M. Jagerschmidt] a préparé l'amendement suivant qui prendrait place entre les articles 54 et 55 : "Dans le cas où une difficulté se produirait sur l'une des dispositions principales de la Convention, une commission spéciale, composée des délégués des différents États, se réunira, sur la demande de l'un d'eux, dans la capitale où aura lieu la dernière Conférence, pour fixer le sens précis à donner à cette disposition. Les décisions de cette commission seront obligatoires pour ceux des États qui n'auront pas cru devoir s'y faire représenter". M. Jagerschmidt développe les motifs de cet amendement. Des divergences d'opinion peuvent se produire sur le sens exact de quelques dispositions de la Convention. Or, l'expérience démontre qu'en pareil cas l'accord s'établit difficilement par correspondance, et que les discussions en conférence conduisent à des résultats plus prompts et plus décisifs. Si donc une de ces difficultés se présentait, il y aurait intérêt à pouvoir la faire résoudre par une Commission. Cette Commission, convoquée spécialement sur la demande d'une ou de plusieurs Administrations, se réunirait dans la capitale de l'État où la dernière Conférence aurait eu lieu. Elle examinerait les questions douteuses, en donnerait la solution et se séparerait

French proposal at Paris had been rejected because it was not concrete enough[6], and so he illustrated the first amendment in great detail. Firstly he gave priority to obtaining efficient communication between the TU administrations, especially for notice of tariff or technical variations on the international network[7]. Once this had been achieved, he felt that there would be no further chance for any kind of misunderstanding, which he saw as lying at the base of the different interpretations of the Convention, an issue raised by Jagerschmidt[8]. Curchod's proposal was to replace the committee with a special agent, who could be called the *Secrétaire-Général des Conférences,* appointed by the assembly and overseen by the telegraph administration organizer of the last conference[9]. The agent

---

immédiatement après. Son utilité paraît incontestable. Elle ne saurait, d'ailleurs, imposer de dépenses bien considérables aux divers États; car ses réunions seraient rares, et ses travaux de courte durée; les États qui n'auraient pas intérêt à s'y faire représenter s'en dispenseraient: les autres pourraient confier à un même délégué le soin de veiller aux intérêts de plusieurs d'entre eux», *Documents de la Conférence Télégraphique Internationale de Vienne*, 379-380.

[6] «La proposition de la France a échoué, non pas que la Conférence en désapprouvât le but, mais parce que l'organisation de la commission n'était pas suffisamment déterminée», ibid., 382.

[7] «Qu'entre ces vingt-six Administrations il existe dans l'application, sur certains points, des divergences d'opinion ; que les modes de procéder varient parfois ; que les renseignements fournis par chacune des vingt-six Administrations aux vingt-cinq autres ne parviennent pas toujours à toutes leurs adresses ; qu'ils ne soient pas conçus dans des termes parfaitement clairs pour tous ; qu'en un mot il y ait défaut d'unité et hésitation dans l'application, c'est ce qui doit assez naturellement résulter de l'état de choses actuel et ce que l'expérience paraît avoir suffisamment démontré», ibid., 381.

[8] «Les difficultés qui se sont présentées entre Administrations dans l'application de la Convention de Paris, sont résultées essentiellement, si ce n'est uniquement, de malentendus, de renseignements insuffisants ou mal compris. On pourrait citer des cas où pendant des mois, presque des années, deux Administrations ont appliqué réciproquement à leurs correspondances communes, l'une les taxes de la Convention de Paris, l'autre les taxes de la Convention de Bruxelles, et cela de la meilleure foi du monde. Il est clair que des malentendus de ce genre ne peuvent guère avoir lieu qu'entre Administrations éloignées les unes des autres, qui n'ont pas habituellement de communications directes entre elles et qui, s'en tenant à des circulaires qu'elle croient de part et d'autre avoir reçues et comprises, n'ont ni occasion, ni raison déterminante de se demander des explications ? A l'occasion de chaque circulaire qu'on reçoit, et chacune des vingt-cinq Administrations correspondantes? Il n'est pont dit que cela éclaircisse beaucoup les choses», ibid., 383.

[9] «Sous la direction de l'Administration télégraphique de l'Ètat où la dernière Conférence a eu lieu, fonctionnerait, pour tout ce qui concerne le service des

would receive all communications about the state of the network from the individual states and send bulletins around periodically. This would simplify proceedings, since the praxis was for all administrations to give notice of any changes on national territory to all the other member states[10]. Facilitated by being in an intermediary position, the agent could also draw up updated tariff tables with ease[11] and publish regularly statistics on the international service[12]. To avoid adding excessive costs to the single administrations, a new special agent would be elected after every conference and would operate under the direction of the administration which had organized the previous conference[13]. The costs of this body

---

communications et renseignements relatifs à l'exécution de la Convention, un agent spécial, nommé par la Conférence et à la solde commune de tous les Ètats contractants. Cet agent, auquel on donnerait par exemple le titre de Secrétaire-générale des Conférences, remplacerait pour les affaires courantes et de peu d'importance la Commission dont il vient d'être question, Commission que rien n'empêcherait cependant de constituer aussi, mais pour n'être appelée à siéger que dans les cas peu probable où une difficulté sérieuse à se présenter», ibid., 282.

[10] «Chaque Administration lui adresserait par simple lettre, aussi souvent que cela lui conviendrait, les communications dont il s'agit ; et périodiquement, sous une forme connue et toujours la même, l'agence expédierait une circulaire fournissant tous les renseignements qu'elle aurait reçus. Au lieu des six cent cinquante documents échangés, il y aurait vingt-six lettres adressées à l'agence et une circulaire de cette dernière à vingt-six destinations, soit en tout cinquante-deux documents», ibid., 384.

[11] «L'agence pourrait même aller plus loin en fait de tarifs. [...] Il serait sans doute difficile, mais non point impossible, d'arriver à l'établissement de tarifs généraux qui déchargeraient toutes les Administrations du travail que leur donne aujourd'hui la confection de leurs tarifs et leur procurerait des tarifs à la fois plus complets et plus économiques», ibid.

[12] «Et la statistique si instructive, si utile en maintes circonstances! Les documents trop peu nombreux que les offices échangent à ce sujet, son conçus sous des formes si diverses qu'il est très difficile d'en tirer parti. Une formule de statistique étant adoptée d'un commun accord par décision de la Conférence ou par l'intermédiaire de l'Administration directrice, l'agence centrale ferait les démarches nécessaires pour qu'elle soit remplie par chaque Administration, et publierait les tableaux qui en résulteraient», ibid., 385.

[13] «L'Administration directrice se chargerait sans doute de ce service; mais par cette combinaison, le but de l'institution ne serait pas atteint complètement: En premier lieu, conformément à l'article 56 de la convention, des Conférences auront lieu successivement dans la capitale de chacun des États contractants, entre les délégués des dits États, et par conséquent tous les deux ou trois ans ce service changerait de main ; son organisation serait en même temps plus ou moins modifiée et il n'y aurait plus ni suite, ni système. En second lieu, ce service qui est dans l'intérêt de tous, serait à charge d'un seul, ce qui est peu naturel et pourrait

would be shared equally among the various states, bearing in mind their dimensions. Nevertheless, they would not be high since they would cover only the agent's salary, the running of the office and travel expenses[14]. The project had been personally drawn up by Curchod and had already received the full approval of the Swiss minister and government. Therefore, though he did not hold it binding, Curchod asked the assembly to approve his proposal as it was, so that he could respect his government's mandate[15].

Whatever, a few changes were made, which turned out to be in Switzerland's favour. The first was proposed by Vinchent, who gave a profile of the ideal agent as a skilled speaker of foreign languages with a profound knowledge of the international telegraph service. For these reasons he felt it essential to fix a permanent and not movable seat for the agent to work from[16]. Vinchent also suggested extending the Bureau's competences to dealing with doubts in interpreting the Convention and possible proposals for altering rules and tariffs[17]. The Prussian delegate Chauvin added a further need, that of locating the seat in a neutral town. In addition, taking his cue from Vinchent he observed that the agent should

---

devenir très onéreux pour certaines Administrations ; enfin, ceux qui auront besoin de recourir à ce service, pourront le faire beaucoup plus commodément et avec plus d'insistance auprès d'une Administrations qu'ils craindraient d'importuner», ibid.

[14] «Les frais de cette agence seraient supportés par toutes les Administrations ils seraient pour chacune d'elles en rapport avec son importance. D'ailleurs, ces frais ne seraient pas bien élevés ; il s'agirait du traitement du Secrétaire-général, puis dépenses de bureau et de voyages; 25 ou 30 mille francs par an suffiraient», ibid., 385-386.

[15] «M. Curchod termine en déclarant que, si son Gouvernement attache un grand intérêt à faire prévaloir le principe de son amendement, il ne refusera aucune concession de forme dont l'utilité serait démontrée», ibid., 387.

[16] «Une pareille mission exigerait, pour être convenablement remplie, un homme joignant l'expérience du service télégraphique et des affaires à un mérite reconnu, connaissant même plusieurs langues. Mais on le trouverait difficilement si l'obligation de changer de résidence tous les trois ans et de se transporter de capitale en capitale lui était imposée. La fixité de la résidence est condition essentielle d'un bon choix», ibid., 388.

[17] «L'Agent-général correspondrait directement avec toutes les Administrations télégraphiques, leur fournissant les renseignements, les explications dont elles auraient besoin et ne réclamant le concours de l'office Directeur que lorsqu'une décision serait à prendre. Il y aurait lieu de recourir à ce concours, quand il s'agirait de modifications au Règlement ou d'interprétations de la Convention», ibid.

not be changed at the end of every conference, but needed to be someone they could count on for many years[18].

Jagerschmidt, the author of the first amendment, was against an "international agent" in the terms defined by Curchod, Vinchent and Chauvin, and accused them of being illiberal, given that the creation of a permanent body of that type would curtail the freedom of decision and action of the single administrations[19]. But as usually happened in the period, the defenders of Curchod's proposal also invoked liberalism, saying that the agent's vigilance and surveillance would guarantee the freedom of the administrations, especially if the agent was a "a man highly respected for his character and personal value"[20]. The discussion lit up on the presumed liberal/illiberal nature of the new body, until the Dutch delegate Staring proposed an article of mediation. The convention would provide for the institution of an office by one of the administrations, which would deal with the tasks of the agent proposed by Curchod. In this way:

[18] «Quant au siège de l'Agence générale, M. le Colonel de Chauvin proposerait de le placer dans une ville neutre. Mais ce à quoi il tiendrait avant tout, ce serait à éviter les changements de personne, qui font perdre au service le fruit de l'expérience acquise, en retardent, quelquefois même en troublent la marche. La personne choisie devrait donc l'être dans des conditions qui permissent de compter sur elle pendant de longues années», ibid., 389.

[19] «Cet agent, tel que le conçoit M. Vinchent, se mouvrait dans une sphère plus large, en dehors de tout contrôle. Les travaux de statistique seraient la partie secondaire de sa tâche ; il serait surtout le conseiller des Administrations, le régulateur du service international. M. Jagerschmidt repousse une pareille situation, qui lui paraît incompatible avec la dignité et la liberté des Administrations télégraphiques. L'agent Général, d'autant plus indépendant des divers offices que l'unanimité de la Conférence, nécessaire pour le nommer, le serait également pour lui retirer ses fonctions, ne tarderait pas à devenir un obstacle pour le service international au lieu d'être une force pour lui», ibid., 390.

[20] «M. Vinchent répond que, l'Agent-général n'ayant aucune décision à prendre, aucun moyen d'imposer ses avis, la liberté des Administrations demeurerait entière. Il exercerait une influence purement morale, mais d'autant plus grande qu'il se recommanderait d'avantage, par l'autorité de sa position, son expérience et sa capacité. C'est pour cela que M. Vinchent voudrait donner une importance véritable à l'institution nouvelle, la constituer indépendante et la rehausser par le choix d'un homme honoré à cause de son caractère et de sa valeur personnelle», ibid., 391.

> It would no longer be the question of a secretary general, which divides opinions, added M. Staring. The International Bureau would have stability and would gather together all the advantages of a secretary general[21]

Faced with a more conciliatory formulation, the French delegation supported it, partly because Curchod had wisely stressed how akin the French proposal was to his. The delegates approved both the establishment of the Bureau and a special committee, which, as in the French proposal, was to be composed of members of all the different administrations convened to judge and settle possible objections about the convention and the rules[22]. The committee was however *de facto* to have a short and unhappy life, as it only met once, in September 1871, to resolve a question about the tariffs of extra-European telegrams to British India. Because of its disappointing results the delegates brought up the subject of the role and functions of the special committee, and after a heated debate[23] decided to abolish it[24].

The *Bureau International des Administrations Télégraphique* was officially named such when article 61 was definitively formulated to provide for its institution and regulations[25]. It was only in the last sitting,

---

[21] «Il ne serait plus question du Secrétaire General sur lequel les avis sont partagés, ajoute M. Staring ; le Bureau international aurait de la stabilité et réunirait tous les avantages qu'on avait demandés à l'institution d'un Secrétaire général», ibid., 394.

[22] «Dans le cas où une difficulté se produiraient sur l'interprétation de l'une des dispositions de la Convention, l'Administration de l'État où aura eu lieu la dernière Conférence convoquera, sur la demande d'une o de plusieurs Administrations, une Commission spéciale composée des délégués des États contractants, et désignera le lieu de la réunion», art. 60, *Convention Télégraphique International de Vienne 1868*.

[23] *Documents de la Conférence Télégraphique International de Rome*, 471-478.

[24] Ibid.

[25] «Une administration télégraphique, désignée par la Conférence, prendra les mesures propres à faciliter, dans un intérêt commun, l'exécution et l'application de la Convention. A cet effet elle organisera, sous le titre de "Bureau international des Administrations télégraphiques", un service spécial qui fonctionnera sous sa direction, dont les frais seront supportés par toutes les Administrations des États contractants et dont les attributions sont déterminées ainsi qu'il suit : Il centralisera les renseignements de toute nature relatifs à la télégraphie internationale, rédigera le tarif, dressera une statistique générale, procédera aux études d'utilité commune dont il serait saisi, et rédigera un journal télégraphique en langue française. Ces documents seront distribués par ses soins aux offices des États contractants. Il instruira les demandes de modification au Règlement de service et, après avoir obtenu l'assentiment unanime des Administrations, fera promulguer en temps utile les changements adoptés», art. 61, *Convention Télégraphique Internationale de*

on 19 July 1868, that the running of the Bureau was officially given to Switzerland[26]. In spite of the fact that the head of the Bureau was to be appointed directly by the administration in command, which in this case was the Swiss one, Chauvin deliberately proposed nominating Curchod in advance and also specified the salary he should receive[27].

The institution of the Bureau was first of all a great success for Switzerland, just as it was for Louis Curchod[28], and thanks to this operation the TU was the first of numerous international organizations to establish its headquarters in the country. Not all plain sailing, however, because at Rome Meydam and the German delegation tried to modify the article dealing with the Bureau in a blatant attempt to snaffle it from Swiss control. Nevertheless, once again Curchod, present in his dual role as representative of a private company and caretaker Swiss delegate, managed to defend the Swiss position of privilege by referring to the results of the first three years[29].

Indeed, as a total innovation both in the fields of international organizations and telegraphic multilateralism, the creation of the Bureau was also a great success for the TU itself.

During the project phase Curchod had been largely inspired by the Swiss political institutions, in particular the Federal Chancellery in the

---

Vienne 1868. For the draft see : *Documents de la Conférence Télégraphique International de Vienne*, 433-434.

[26] «La Conférence procédant au choix de l'office auprès duquel sera installé le bureau international désigne à l'unanimité l'office Suisse. M. Le Colonel de Chauvin désirerait que la Conférence émit le vœu formel de voir M. Curchod placé à la tète de ce bureau», ibid., 454.

[27] «Il [M. Chauvin] voudrait, au contraire, faciliter l'exercice de son droit et aller au devant de certains scrupules que le Conseil fédéral éprouvera certainement, en lui donnant l'assurance que, si son libre choix tombait sur M. Curchod, la Conférence, qui a pu apprécier tout ce qu'il ya chez lui, de savoir et d'expérience, y verrait un gage d'avenir pour l'institution du Bureau International, qu'il considère comme le résultat le plus important des travaux de la Conférence ; en lui faisant savoir, en outre, que les services de Directeur de ce bureau ne lui paraîtraient pas trop largement rémunérés par un traitement annuel de douze mille francs», ibid., 456.

[28] Gabriele Balbi, Simone Fari, Giuseppe Richeri, Spartaco Calvo, "Swiss specialties. Switzerland's role in the genesis of the Telegraph Union," *Journal of European Integration History* 19, no 2 (2013): 207-225.

[29] *Documents de la Conférence Télégraphique International de Rome*, 481-485. Curchod was actually Head of the Bureau until 1869, when he stepped down in order to take up a managerial position with the *Société du câble transatlantique*. Lendi filled in until his death 1872, when Curchod took over as Head of the Bureau again.

period of the Vorort government[30]. And during the revision and discussion phase at the Vienna conference Vinchent was influenced by some of the agencies that ran cross-border infrastructures of communication like l'*Agent Général des chemins de fer Rhénans de l'État Belge et de la Compagnie française du Nord*[31].

That those involved were aware they had obtained a great result also emerges from the way felicitous expressions they later adopted, for example Brunner in the 1871 Rome conference:

> M. Brunner regards the creation of the International Bureau as one of the happiest ideas in the Vienna Conference and one of the more efficacious guarantees of the union among states. The aim of this institution is to facilitate relations between governments and remove any difficulties that could arise. All administrations have profited from its works on statistics and tariffs and the Bureau will certainly be able to satisfy the idea which has come of making a central body for the technical part of the telegraph service[32].

---

[30] «[...] dans la Confédération Suisse, il y a 25 ans, la direction des affaires fédérales était entre les mains du Gouvernement d'un Canton et passait tout les deux ans d'un Gouvernement à l'autre. *Le Gouvernement changeait ainsi que le lieu de résidence, mais ce qui ne changeait pas c'était la Chancellerie fédérale, armée de ses Archivees, de ses documents et de ses traditions,* qui préparait le travail du Gouvernement du Vorort et mettait ses décisions a exécution. Grâce à la Chancellerie fédérale ces changements réguliers de Gouvernement se faisaient sans choc, sans arrêt des affaires sans compromettre l'unité du système et les gouvernements des Cantons directeurs pouvaient d'un autre coté se charger successivement de la direction des affaires fédérales sans s'exposer à des sacrifices sensibles, car les frais de la Chancellerie étaient couverts par l'ensemble des Cantons», Archive Fédérale (AF), Fond E52, Archive nr. 509, *Louis Curchod à Jean-Jaques Challet-Venel*, 12 mars 1868.

[31] «M. Vinchent tient à expliquer comment il entendrait l'institution nouvelle proposée par la Suisse ; car, restreinte dans les limites définies par l'amendement, elle ne rendrait pas tous les services qu'il voudrait lui demander. L'institution, telle qu'il la comprend, fonctionne déjà dans la pratique. Sous le titre d'Agent-général des chemins de fer Rhénans, de l'État Belge et de la Compagnie française du Nord, un agent résidant à Cologne fournit à chacune de ces trois Administrations les renseignements dont elle a besoin, s'occupe des affaires communes à leurs réseaux, et en prépare les solutions ; en cas de désaccord, il donne son avis, mais sans jamais prendre aucune décision. C'est un rôle analogue que devrait remplir le nouvel agent télégraphique», *Documents de la Conférence Télégraphique International de Vienne*, 388.

[32] «M. Brunner regarde la création du Bureau international comme une des idées plus heureuses de la Conférence de Vienne et une des garanties les plus efficaces de l'union des États. Cette institution a pour but de faciliter les rapports des

And when four years later at St Petersburg Curchod was awarded an important decoration by the Russian government, he declared that one of his greatest merits was having promoted the cause of the Bureau at the Vienna Conference[33].

3.2 The functions and powers of the Bureau

According to the post 1868 conventions and rules, the Bureau was mandated to carry out five main tasks: 1) collect information; 2) process, edit and publish the said information in works or reviews; 3) send official communications to member states; 4) archive correspondence about suggested changes to rules and conventions; 5) offer consultancy for technical aspects of the international service[34].

Gouvernements entre eux et de faire disparaître les difficultés qui peuvent surgir. Toutes les Administrations ont profité déjà de ses travaux sur la statistique et sur les tarifs, et elle sera certainement en mesure de donner satisfaction à l'idée qui s'est produite d'en faire un organe central pour la partie technique du service télégraphique» *Documents de la Conférence Télégraphique International de Rome*, 484.

[33] «M. Curchod considère comme un de ses titres les plus honorables d'en avoir été le promoteur à Vienne et d'avoir contribué à la fonder. Ses fonctions le mettent en rapport avec toutes les Administrations et spécialement avec l'Office qui est chargé de diriger les travaux préparatoires des Conférences», *Documents de la Conférence Télégraphique International de St. Petersburg*, 650.

[34] Art. 61, *Convention Télégraphique Internationale de Vienne 1868*. «Les offices des États contractants se transmettent réciproquement tous les documents relatifs à leur administration intérieure et se communiquent tout perfectionnement qu'ils viendraient à y introduire. En règle générale le Bureau international sert d'intermédiaire à ces notifications ; toutefois les avis à transmettre d'urgence, et spécialement la notification des interruptions de lignes, son directement portés par la voie télégraphique à la connaissance de toutes les Administrations intéressées. Les dites Administrations envoient par la poste, par lettre affranchie, au Bureau international, la notification de toutes les mesures relatives à la composition et aux changements des tarifs, tant intérieurs qu'internationaux, à l'ouverture de lignes nouvelles et à la suppression de lignes existantes en tant que ces lignes intéressent le service international ; enfin aux ouvertures, suppressions et modifications de service des bureaux. Elles lui font parvenir au commencement de chaque année, et aussi complètement qu'il leur est possible, des tableaux statistiques du mouvement des correspondances, de la situation des lignes, du nombre des bureaux et des appareils, etc. Ces tableaux son dressés d'après les indications du Bureau international qui distribue à cet effet les formules toutes préparées. Elles adressent également à ce bureau deux exemplaires des publications diverses, qu'elles font paraître. Le Bureau international reçoit en outre communication de tous les

On paper and in the intentions of various administrations these tasks were to confine the Bureau to a purely administrative role. Yet its intermediary position, indispensable for carrying out its prescribed functions, guaranteed it the privilege of direct contact with all the administrations, and consequently find itself well positioned to influence either directly or indirectly individual choices. As we have seen, Curchod's designs for the Bureau were born of a need to improve communications between the member states, so clearly intermediation would be right from the beginning the Bureau's main activity, as Curchod himself declared[35].

---

renseignements relatifs aux expériences auxquelles chaque Administration a pu procéder sur les différents parties du service», art. XXXII, *Règlement de service international, Convention Télégraphique International de Vienne.* «Indépendamment des communications spéciales que le Bureau international est tenu de faire à toutes les Administrations, il utilise les documents de statistique et autres qui sont mis à sa disposition, pour la rédaction du journal don il est fait mention à l'article 61. Il doit d'ailleurs se tenir en tout temps à la disposition des Administrations des États contractants pour leur fournir, sur les questions qui intéressent la télégraphie internationale, les renseignements spéciaux de tous genres dont elles pourraient avoir besoin.», art. XXXIII, *Règlement de service international, Convention Télégraphique International de Vienne.* «[…] 2. Il dresse, publie et révise périodiquement la carte officielle des relations télégraphiques. […] 4. Dans les questions à résoudre par l'assentiment des Administrations contractantes, celles qui n'ont point fait parvenir leur réponse dans le délai maximum de quatre mois, son considérées comme consentantes. 5. Les documents imprimés par le Bureau international sont distribués aux Administrations des États contractants dans la proportion du nombre d'unités contributives, d'après les articles XXXII et XXXV. Les documents supplémentaires que réclameraient ces Administrations son payés à part, d'après leur prix de revient. Il en est de même des documents demandés par les exploitations privées. 6. Les demandes de cette nature doivent êtres formulées une fois pour toutes, jusqu'à nouvel avis, et de manière à donner au Bureau international le temps de régler le tirage en conséquence. 7. Le bureau international prépare les travaux des Conférences télégraphiques. Il pourvoit aux copies et impressions nécessaires à la rédaction et à la distribution des amendements, procès verbaux et autres renseignements. 8. Le directeur de ce bureau assiste aux séances de la Conférence et prend part aux discussions sans voix délibérative.», art. XXIV, *Règlement de service international, Convention Télégraphique International de Rome.*

[35] «Le premier soin du bureau international a été de se mettre en relation avec les Administrations de tous les États contractants et de toutes les Compagnies dont les réseaux se relient directement à ceux de ces États», ITU-Corr., feuille n. 1/121, 1869, *Rapport de gestion du Bureau International des Administrations Télégraphiques Année 1869.*

Above all, in collecting information on office opening and closing times, the inauguration/interruption of landlines or circuits and the possible breakdowns/repairs of submarine cables, the Bureau was monitoring constantly and totally the international network. In doing so it was able to carry out two other important tasks: signal periodically any changes to all administrations and put them together in official publications.

Once the Bureau had collected all the reports of variations in the international network, it circulated them periodically in communiqués known as "notifications"[36]. Another task was to collect all data about variations in the network into publications: the opening/closing of offices in the *Nomenclature*, notice of cable laying/breakdowns/repairs and new lines opening in the TU's official review *Journal Télégraphique* (JT), while statistics (length of lines and wires, traffic, personnel) went in the *Statistique Générale*, also included in the JT.

These operations of collecting and circulating information either directly or indirectly did not by themselves influence in any great way the administrations because they were simply exchanges of technical news on the functioning of the service. They formed, though, an intense epistolary exchange, which served to strengthen a sense of trust between the administrations and the Bureau. Where the Bureau was able to exert a greater influence was through: 1) the publication of the official review; 2) general consultancy on telegraph questions; 3) the collection of proposals for modifying conventions and rules.

The *Journal Télégraphique* consisted of the contributions sent by administrations to the Bureau, which then made the selection for publication. Already in this way the Bureau was able to exert a notable power, further increased by its right to edit the articles, which came out as in-house publications. Most of the Bureau's articles were sourced from reports on specific matters regarding the telegraph service, which could originate in some question from an administration or directly on the initiative of the Bureau itself. For example, in 1869 the Bureau wanted to know the level of female employment and sent around a circular to all administrations asking for data[37]. Within a few months all the administrations had answered[38] and the Bureau was able to publish its

[36] The *notifications* were defined as «ordinairement mensuelles, par lesquelles sont portés à la connaissance des offices de l'Union les renseignements de tous genres concernant le service télégraphique international», *L'Union Télégraphique Internationale (1865-1915)*, 34.

[37] ITU-Corr., feuille n. 90/3, 1869.

[38] ITU-Corr., Dossier n. 90, 1869.

results in two issues of the *Journal Télégraphique*[39]. Two years later the Bureau published in the same journal a report[40] into the occupational pensions of telegraph personnel[41], which had been sparked off by a request for information from the head of Italian telegraphs, that because of an impending reform in Italy. Furthermore, as the Bureau published the *Journal Télégraphique*, it is most likely to have used it as its preferred channel for circulating its own opinions on the international service. From this point of view even the apparently quite boring correspondence about subscriptions or mailings of copies of the *Journal Télégraphique* or other official Bureau publications can turn out to convey more than simple administrative operations. The capillary circulation of official publications met the double target of making the telegraphic service more functional and passing on to the administrations the opinions of the TU, i.e. those of the Bureau head[42].

The Bureau was able to exercise even greater influence via the technical consultancy service for individual administrations, which turned to it whenever they wanted information on some specific aspect of the telegraph service. If the matter was of general interest, as in the case of the pensions, a report was carried out and published in article form in the *Journal Télégraphique*. Sometimes, however, the information was deemed

---

[39] "De la participation des femmes au service télégraphique (1ere article) ," *Journal Télégraphique* 2 (1869) : 9-11; "De la participation des femmes au service télégraphique (2ème article)", *Journal Télégraphique* 3 (1870) : 21-27.

[40] ITU-Corr., Dossier n. 83, 1871.

[41] "Des pensions de retraite dans le service télégraphique (1ère article) ," *Journal Télégraphique* 24 (1871): 361-364; "Des pensions de retraite dans le service télégraphique (2ème article)," *Journal Télégraphique* 25 (1871): 377-381; "Des pensions de retraite dans le service télégraphique (3ème article)," *Journal Télégraphique* 1 (1872): 1-3; "Des pensions de retraite dans le service télégraphique (4ème article)," *Journal Télégraphique* 2 (1872): 17-19; "Des pensions de retraite dans le service télégraphique (5ème article)," *Journal Télégraphique* 3 (1872): 33-38; "Des pensions de retraite dans le service télégraphique (6ème article)," *Journal Télégraphique* 4 (1872): 49-51; "Des pensions de retraite dans le service télégraphique (7ème article)," *Journal Télégraphique* 5 (1872): 65-68; "Des pensions de retraite dans le service télégraphique (8ème article)," *Journal Télégraphique* 6 (1872): 81-85; "Des pensions de retraite dans le service télégraphique (9ème article)," *Journal Télégraphique* 7 (1872): 97-99; "Des pensions de retraite dans le service télégraphique (10ème article)," *Journal Télégraphique* 8 (1872): 113-118; "Des pensions de retraite dans le service télégraphique *(11ème article)*," *Journal Télégraphique* 9 (1872): 129-133.

[42] ITU-Corr.

unpublishable. For example, again in 1869, the head of Italian telegraphs, D'Amico, set up an enquiry into the health of the telegraph operators[43] since he had noticed from local statistics that pneumonia was the major cause of death among operators. In order to establish whether there was any correlation between lung complaints and working with telegraph instruments, D'Amico asked the Bureau to investigate among the other administrations[44]. The results of the ensuing report[45] were circulated in a confidential way. Once the data had been analyzed and summarized in a general table, the Bureau informed the Italian administration of its findings[46]. Lendi, then head of the Bureau, felt that the high rate of deaths could not be seen as an occupational hazard and quoted as proof the Belgian statistics that pulmonary illnesses were frequent in public administration staff in general. He was also convinced that as telegraph operating was a sedentary activity, it attracted those already in poor health.

What was clear was that the Bureau had no intention of going further into the matter, and even less of taking action. It only reminded the administrations that office space had to be well aired and that the batteries had to be stored apart, outside the work environment. In this case, indeed, the Bureau's opinion functioned as a directive and all the administrations ended by thinking like the Bureau and judging the available statistics insufficient. Thus the question of any relation between lung diseases and working on telegraphic machines was shelved.

If other reports mandated by the Bureau are analyzed, it can be seen that it exerted a far greater influence over administration decisions than delegates had initially ever imagined. Whenever an enquiry was opened to obtain information from the administrations or attempt to judge the opportunity of a project, the Bureau sent around a circular which was very rarely a straight request for information. It was very often divided into two parts, the first inviting the administrations to give their opinion about something, the second supplying them with an idea of what the Bureau thought. A telling example is given by the correspondence over establishing an international school[47]. In July 1872 the Italian administration informed the Bureau of its idea to set up an international institute in order to guarantee telegraph staff a high level of training[48]. The Bureau at once sent around a circular inviting administrations to express

[43] ITU-Corr., dossier n. 74, 1870.
[44] ITU-Corr., feuille n. 74/1, 1870.
[45] ITU-Corr., feuille n. 74/2, 1870.
[46] ITU-Corr., feuille n. 74/19,1870.
[47] ITU-Corr., dossier n.132, 1872.
[48] ITU-Corr., feuille n. 132/1, 1872.

their opinion[49] but adding its own ideas on the project. And Lendi, as Bureau head, felt the idea was impractical, for the difficulty in finding a common language in which to teach and recruiting sufficiently skilled teachers, as well as the problem of the diverse technical level of the various networks[50]. In brief, the Bureau rejected the idea, and the effects of its judgement were inevitably felt on the decisions of the administrations. Most of them in fact gave a negative opinion and put forward the same reasons as the Bureau. All this goes to show that the Bureau did not exert influence only in exceptional circumstances, as Codding holds, but it was routine practice, reiterated as far as possible as the starting point of every enquiry[51].

Lastly, the Bureau was able to exert its greatest influence in the procedure for modifying rules and tariffs. In theory, it collected and circulated the proposals from one state to all the others, and then awaited their opinions. In practice the Bureau rarely lay low, and included its own comments with the proposals, or conversely the administrations solicited its opinions. The case in which its power emerged most evidently was in the procedure for drafting the St Petersburg Convention. The Rome Conference had decided to entrust the Bureau with an important task, i.e. prepare a: "*modification* of the Convention and Regulations which must be communicated to all the Offices about a year before the next Conference"[52]. In other words, the head of the Bureau would be called on to work on the text of a proper constitution for the Union, based on the Rome version, and as we have seen, the Convention, Regulations or Minutes of the Rome Conference did not make it clear if the new convention project was to be masterminded or simply assembled by the Bureau. In a literal interpretation, the Bureau was to limit itself to collecting and publishing proposals coming from administrations belonging to the Union. Furthermore, following the custom of the previous conferences, the creative and proactive role in drawing up the Convention was to come from the administration organizing the conference, in this case the Russian one. This ambiguity emerged and was resolved in a correspondence between the Bureau and the Russian administration over an apparently different topic: the voting system to adopt. The exchange of letters also clarifies how important the Bureau and its head had become. It

---

[49] ITU-Corr., feuille n. 132/3, 1872.

[50] Ibid.

[51] Codding, *The International Telecommunication Union*, 51.

[52] «Remaniement de la Convention et du Règlement qui devrait être communiqué à tous les Offices environ une année avant la prochaine Conférence». *Documents de la conférence télégraphique internationale de Rome*, 443.

was started by Lüders, the head of Russian Telegraphs, to the Bureau, bringing up a question which had been left unresolved in Rome, i.e. "if the vote should belong to the States in the diplomatic sense of the word or to the Telegraph Administrations"[53].

As organizer of the coming Conference, the Russian administration was to assume the chair and as such would have to decide if to allow the issue of the colonial vote to be aired afresh. Lüders then asked the Bureau to set in motion an opinion poll among other administrations in order to find out what they thought about the voting procedure, a task which was carried out by the secretary De St-Martial, acting head of the Bureau[54]. The result was, as we have seen, that most member administrations were against the multiple vote and those in favour like Spain and Italy hoped to gain advantages for their countries. Whatever, Curchod had in the meantime resumed the role of head and took the project of the new convention firmly into his hands by means of the survey results and a dense correspondence.

In summer 1873, when the survey was coming near to its end, Curchod decided to communicate the provisory results to the head of Russian Telegraphs in a somewhat unusual way, sending him two letters on the same day. The first had an official character and reported in a neutral professional way the opinions of all the administrations, without ever quoting Curchod's[55]. The second was confidential, a kind of private communication between Curchod and Lüders[56], as the former carefully anticipated[57]. More interesting than the first letter, the second can be read as Curchod's covert attempt to influence the preparatory stage of the conference as he moved and manoeuvred moving along the line of a normative vacuum. He knew that the preceding documents did not attribute either to the Bureau or the Russian Administration the task of rewriting the new convention, and therefore made a successful attempt to turn the tables in his favour. He declared quite openly that the voting

---

53 «le vote devait appartenir aux Ètats dans le sens diplomatique du mot ou bien aux Administrations télégraphiques», *Lettre écrite par l'Administration Russe adressée au Bureau International*, 18 Décembre 1872, in ITU-Corr, feuille n. 29/1, 1872.

54 *Lettre écrite par le Bureau International adressée à toutes les Administrations*, 10 Janvier 1873, in ITU-Corr, feuille n. 29/2, 1873.

55 *Lettre écrite par le Bureau International adressée à l'Administration Russe*, 11 Juillet 1873, in ITU-Corr, feuille n. 29/20, 1873.

56 *Lettre écrite par le Bureau International adressée à l'Administration Russe (confidentielle)*, 11 Juillet 1873, in ITU-Corr, feuille n. 29/22, 1873.

57 «ne peut avoir, d'ailleurs [...] qu'un caractère tout personnel de ma part», ibid.

question was not in itself very important, almost as though to play down the emphasis Lüders was giving it. Nevertheless, he wanted to furnish a reply, and taking the cue from Vinchent, his historical ally/adversary, he declared his opposition to a multiple vote. He was in favour of keeping one vote, one state, and left open the question of giving the vote to multiple administrations within a state. Nevertheless, probably because he was reluctant to oppose Lüders and wanted to show yet again his level of professional correctness, he suggested at the end of his letter a simulation of a multiple vote[58]. With Spain, Italy and Russia's proposals in mind, Curchod formulated a series of indicators (e.g. the number of inhabitants and territory size) with which to allocate the number of votes to the different states. The aim was obviously to give a scientific backing to his opinion, but in such a way as not to appear factious or presumptuous in Lüders' eyes.

The real reason for the confidential letter to Lüders was not so much to fix the voting question but to let him know what he thought about the future development of the Union. Curchod began to go into the matter by commenting on the Italian's observations on the multiple vote and his emphasis on the fact that before discussing voting procedure they needed to know what type of document would be approved in the St Petersburg and successive conferences. D'Amico was in fact already drawing a distinction between diplomatic conferences with a one vote/one state rule and administrative ones, where various administrations within the same state would all have the right to vote. Starting from this position Curchod illustrated his plan for St Petersburg and successive conferences:

> I think that the Paris Convention should be replaced at St Petersburg by a new international treaty brought about through diplomatic channels. This treaty would only contain the basic principles which have been the rule since the Paris Conference and even before and which have never given rise to any serious debate. It will be decided upon and signed by all those who would like to take part with the conditions which I feel must have to put aside, giving voice more to some than to others; in the heart of the diplomatic conference there would be no real vote as such, a complete agreement is necessary but while respecting diplomatic usage, this diplomatic conference could perfectly well delegate to administrative conferences the job of formulating, pending government approval, all

[58] The term simulation is used here in its technical-mathematical meaning. Curchod simulated the application of a new regulation by making a series of calculations which brought to an understanding of how a multiple vote would be divided among the various member states. He used mathematical simulation very much as happens nowadays.

> kinds of measures (regulations and tariffs) needed to implement the conventions. In delegating this task, it could without doubt organize administrative conferences and in particular the way of voting which will have to be respected and that will be able to, of common agreement, hold account in the repartition of ways the number of distinct administrations and their relative importance [...] To sum up, on one side an international treaty, unchangeable until notice of termination and on the other, tariffs which lend themselves to necessary changes and are submitted to periodic reviews according to the forms dictated by the treaty itself.[59]

Thus for the first time Curchod explained his vision of the organization of the Union from St Petersburg onwards. He spoke for the first time of convening a last conference in which diplomats in their role of political representatives of national governments would sign an unchangeable convention. For the first time he mentioned the fact that St Petersburg would be followed by administrative conferences alone.

While the Bureau had been given the task of drawing up a convention in Rome, the idea of making a permanent change to the nature of the conference was aired for the first time in this letter. Curchod had a very clear idea of the evolution of international conferences and he was equally aware that he could not impose it on Lüders and would have to negotiate. The following passage clarifies the point and highlights his diplomatic skills:

---

[59] «Je pense que la Convention de Paris devrait être remplacée à St. Petersburg par un nouveau traité international conclu par la voie et dans la forme diplomatiques. Ce traité ne contiendrait que les principes fondamentaux qui ont fait règle depuis la conférence de Paris et même auparavant, et qui n'ont jamais donné lieu à aucune discussion sérieuse. Il serait arrêté et signé par tous les États qui voudraient y prendre part dans des conditions qui me paraissent devoir écarter toute idée de donner plus de voix aux uns qu'aux autres; dans le sein de la Conférence diplomatique il n`y aurait pas de vote proprement dit, il y faut un accord complet mais cette Conférence diplomatique, tout en respectant les usages de la diplomatie pourrait parfaitement déléguer à des conférences administratives le soin d'arrêter, sous réserve de l'approbations des Gouvernements respectifs, les mesures de tous genres (règlement et tarif) nécessaires à l'exécution de la Convention, et en délégant ce soin elle pourrait aussi, sans aucun doute, organiser des Conférences administratives et en particulier le mode de votations qui devra y être observé et qui pourra, d'un commun accord, tenir équitablement compte, dans la répartitions des voix du nombre des administrations distinctes et de leur importance relative. [...] En résumé, d'une part un traité international immuable jusqu'à dénonciation d'autre part les règlements et les tarifs se prêtent sans cesse aux changements nécessaires et soumises à des révisions périodiques selon les formes dictées par le traité lui-même», ibid.

> I think I must communicate to you with this appendix the convention project I have drawn up, to put in order my ideas, in applying preceding principles and taking as my base the Paris Convention revised at Rome. In my eyes this project is what should have been in the diplomatic convention, if it had been concluded at Rome, having found its place in the regulations and tariffs. I do not for the moment insist any more on this project; I have said enough so that you can […] form an opinion and I beg you let me know as soon as possible. The work on certification prescribed by the Rome Conference can only be usefully undertaken when you have drawn up the plan. In effect, it is on your Department that the direction of operations for the St Petersburg Conference falls, and I cannot and must not be more than an instrument in your hands for the preparatory works.[60]

Curchod was of course well aware he was no simple instrument in Lüders' hands. Quite the reverse, for he attached to the above declaration a project for the convention which not only Lüders would make totally his, but would be discussed and passed without any opposition in St Petersburg. Several studies, in fact, agree on the fact that in the person of Curchod the Bureau reached the apogee of its influence in its preparation of the procedures and contents to be approved at St Petersburg[61].

Though Lüders' answer referred explicitly to Curchod's official letter[62], his words show he must have read the confidential note carefully. In fact, he asked Curchod to explain how to cast his vote in the administrative and diplomatic conferences:

---

[60] «Je crois devoir vous communiqués ci-joint le projet de convention que j'ai préparé, pour me fixer les idées, en application des principes qui précédent et en prenant pour base la convention de Paris révisée à Rome. Ce projet représente donc, à mes yeux ce qui aurait dû figurer dans la convention diplomatique, si on en avait conclu une à Rome, tout le reste ayant trouvé sa place dans le règlement et dans le tarif. Je n'insiste pas pour le moment plus longuement sur ce projet; j'en ci dis assez pour que vous puissiez […] former une opinion et je vous prie de me la faire connaître le plus tôt possibles. Le travail de certification prescrit par la Conférence de Rome ne pourra être utilement entrepris que lorsque vous aurez arrêté le plan. En effet, c'est à votre Département qu'incombe incontestablement la direction des opérations de la Conférence de St. Petersburg et je ne puis ni ne dois être, dans les travaux préparatoires qui la concernent, qu'un instrument entre vos mains», ibid.

[61] Codding, *The International Telecommunication Union*, 51; Fari, *Una penisola in comunicazione*, 473.

[62] *Lettre écrite par le Bureau International adressée à l'Administration Russe*, 11 Juillet 1873, in ITU-Corr, feuille n. 29/20, 1873.

> I beg you, dear Director, to submit again this object to the examination of the Administrations interested and collect their opinions on the decision if the vote should belong to the States in the diplomatic sense of the word or to the administrations, and then give me the results together with your opinion on the subject[63].

It is clear that once Lüders had understood why Curchod judged the multiple vote pointless, he wanted a proposal on how to solve the problem of the colonial vote. He therefore asked Curchod to open a new consultation and then give him the results together with his own opinion. The two heads met in Vienna most probably in August/September 1873[64], where Curchod showed Lüders his project for amending voting procedures: 1) one vote per state for approving the last diplomatic convention; 2) possibility of allocating a vote to various administrations belonging to the same state providing the government in question made an explicit request and the delegates of each administration were present at the conference. An official letter from Curchod to Lüders of 24 October marked the closure of question[65]. Curchod had not only greatly influenced the final version of the voting procedure norm, so crucial for functioning of the conferences, but he had set up the future agenda of the Union and its aim to become a stable structure destined to last in time. What is noticeable is that to reach these aims he deployed the traditional bureaucratic and administrative instruments the conferences had given him. He never formally passed the limits set by convention rules and regulations and, in keeping to them, managed to pilot almost all the most important decisions.

---

[63] «Je vous pries, Monsieur le Directeur, de vouloir bien soumettre de nouveau cet objet à l'examen des Administrations intéressées et en recueillir les opinions sur la décision si le vote devait appartenir aux États dans le sens diplomatique du mot ou bien aux administrations et ensuite me communiquer le résultat simultanément avec votre propre appréciation sur la matière», *Lettre écrite par l'Administration Russe adressée au Bureau International*, 27 Août 1873, in ITU-Corr, feuille n. 29/23, 1873.

[64] «Conformément à ce qui a été convenu récemment entre nous à Vienne, j'ai l'honneur de vous adresser ci-joint, après en avoir revu la rédaction, le projet de vote que j'ai écrit sous votre inspiration et avec votre approbation. Il s'agit plutôt d'une simple esquisse que d'un projet complet, car à supposer même que son contenu fut approuvé et adopté en entier par vous», *Lettre écrite par le Bureau International adressée à l'Administration Russe (confidentielle)*, 28 Septembre 1873, in ITU-Corr, feuille n. 29/25, 1873.

[65] *Lettre écrite par le Bureau International adressée à l'Administration Russe*, 24 Octobre 1873, in ITU-Corr, feuille n. 29/28, 1873.

3. Conclusions

This chapter has highlighted the way the Bureau operated discreetly but at times also disruptively within the TU. Not present at the beginning or in any of the earlier Unions, it was established because of a need for a permanent body to function as an intermediary between administrations during the periods between one conference and another. The driving force behind it was Switzerland, which saw the possibility of exercising a new role as a neutral power on the chessboard of international politics. Greatly undervalued by the delegates who created it, the Bureau finished by playing a political role far above the administrative tasks it had been given. By means of its functions of intermediation and circulation of news, (the *Journal Télégraphique*, official statistics, *Nomenclature*), the Bureau increased its influence right from its beginnings on the creation of norms, up to the point of carrying out a determining role both in and outside the conferences in the construction of the last convention.

CHAPTER FOUR

STANDARDIZATION[1]

Specialist literature has concentrated its studies on the Telegraph Union as a supranational political organization, and generally taken little notice of aspects concerning the standardization of the service[2].

Most particularly, as a first examination of primary sources shows, technical standardization appears to have been considered the least important of the Union's main functions. For example, only two of the 63 articles of the 1865 Paris Convention concerned telegraph technology[3] while most of the articles in the Convention and its Regulations dealt with homogenizing general norms. The delegates appear to have been more at their ease working at standardizing rules for communication rather than delving into the technical means for carrying it out.

The fact that more space and time were taken up with the standardization of norms can be easily explained by the nature of the Union. As we have seen, the delegates drawing up and approving the conventions were mainly the heads of the individual national administrations who dealt with the everyday running of their lines. So it is not surprising, as we shall see, that they often worked by the principles of liberalism dear to the governments they represented, and prioritized the standardization of general norms rather than the materials and technologies needed for transmitting and receiving international telegrams.

---

[1] A first version of the second part of this chapter was presented at Histelcon 2012, "Origins of Electro-technologies", 5-7 September 2012, Pavia, Italy.

[2] Some exceptions in this sense, though not specifically dealing with the TU are: Roland Wenzlhuemer, "The History of Standardization in Europe," *European History Online (EGO)*, Institute of European History (IEG), Mainz 2010-12-03, URL: http://www.ieg-ego-eu/wenzlhuemer-2010-en, [04/05/2014]; Philipp Genschel, Raymund Werle, "From National Hierarchies to International Standardization. Modal Changes in the Governance of Telecommunications," *Journal of Public Policy* 13, no 3 (1993): 203-225; León Alfred Burtz, Eckart Hummel, "Standard Setting in International Telecommunications," *Telecommunications Policy* 8, no 1 (1984): 3-6.

[3] *Convention télégraphique internationale de Paris 1865.*

The causes of the scarce importance attributed to technical standardization are also to be sought in the technological level of the telegraphs of the time, which by the mid-sixties had almost all reached a high level of national standardization, with both the lines and the apparatus tending to converge on the models reputed to be most efficient. Often, therefore, the TU rested on its laurels rather than seek and establish new technological standards. This in spite of the fact that the top men were in favour of developing instruments to favour the birth of a shared managerial culture.

4.1 Standardizing norms

4.1.1 The Liberal Inheritance

As we have seen, the TU's norms were mostly structured on the model of the rules established during the years of its gestation, so it follows that the most important principle inherited from multilateralism was the free circulation of telegrams. Around it the TU carefully and cautiously constructed all its rules and rhetoric.

In a period in which liberalism permeated in theory and practice the political activity of most European governments, international cooperation was seen as an instrument for guaranteeing and expanding the individual freedom of the citizenry[4], which in telegraphic terms was seen in a universal right to correspondence, as declared in the 1865 Convention: "The contracting High Parties recognise all persons' right to correspond by means of International telegraphs"[5]. A freedom which was then strengthened by the equally universal right to privacy for the messages, to be respected and protected not only by the individual operators but also the administrations[6]. Nevertheless, at the same time the liberalism of the period based its strength on and was nourished by the defence of the sovereign state, which inevitably imposed limits on individual freedom[7]. In the same way in which individual initiative and expression of thought

---

[4] Dwayne Winseck, "Globalizing Telecommunications and Media History. Beyond methodological nationalism and the struggle for control model of communication history," in *Global Communication Electric. Business, News and Politics in the World of Telegraphy*, eds. Michaela Hampf and Simone Müller-Pohl (Frankfurt/New York: Campus Verlag, 2013), 44-45.

[5] Art. 4, *Convention télégraphique internationale de Paris 1865.*

[6] «Elles s'engagent à prendre toutes les dispositions nécessaires pour assurer le secret des correspondances et leur bonne expédition», art. 5, *Convention télégraphique internationale de Paris 1865.*

[7] John Stuart Mill, *On Liberty* (Boston: Ticknor and Fields, 1863).

were curbed to protect the community, telegrams could be freely sent as long as there was no intimation of damage to the moral and physical integrity of the state. With this well in mind, the TU recognized the right of administrations not only to violate the privacy of correspondence, but also to block telegrams felt to be dangerous for national security, moral or social order[8]. In addition, international cooperation in the period was openly based on respecting the freedom of action of every national government, so every state could decide in any moment to block international telegrams, subject to notifying other states[9].

The four rights of free circulation, privacy of correspondence, national security and respect of sovereignty so characteristic of liberal ideology, were handed straight down from the older telegraph conventions, to become the basic concepts underlying the Paris Convention and its successors. The norms were so deeply rooted in telegraphic multilateralism that not only were they never queried during the conferences[10], but delegates often quoted them in order to back their amendments or decisions.

Another norm which recapitulated the principles of freedom dealt with the order in which telegrams were to be sent. First to go were state despatches, more precisely "those sent by heads of state, ministers, commanders-in-chief of land or sea forces and diplomats and consuls of the member governments"[11]. They were followed by service telegrams, sent by the administrations concerning the running of the lines[12]. Last in

---

[8] «Les Hautes Parties contractantes se réservent la faculté d'arrêter la transmission de toute dépêche privée qui paraîtrait dangereuse pour la sécurité de l'État, ou qui serait contraire aux lois du pays, à l'ordre public ou aux bonnes mœurs, à charge d'en avertir immédiatement l'expéditeur. Ce contrôle est exercé par les bureaux télégraphiques extrêmes ou intermédiaires, sauf recours à l'Administration centrale, qui prononce sans appel», art. 19, *Convention télégraphique internationale de Paris 1865.*

[9] «Chaque Gouvernement se réserve aussi la faculté de suspendre le service de la télégraphie internationale pour un temps indéterminé, s'il le juge nécessaire, soit d'une manière générale, soit seulement sur certaines lignes et pour certaines natures de correspondances, à charge par lui d'en aviser immédiatement chacun des autres Gouvernements contractants», art. 20, *Convention télégraphique internationale de Paris 1865.*

[10] Even at Paris, the norm was approved without any debate: *Documents Diplomatiques de la Conférence Télégraphique Internationale de Paris*, 117-119.

[11] Art. 7, *Convention télégraphique internationale de Paris 1865.*

[12] «Dépêches de service: celles qui émanent des administrations télégraphiques des États contractants, et qui sont relatives, soit au service de la télégraphie

the pecking order private telegrams. Within a certain category telegrams had to be sent in order of acceptance, to avoid accusations of favouritism and ensure the freedom and equality of all citizens[13]. Once again, the only limits were to give priority to national interests or technical needs guaranteeing the execution of the telegram itself[14]. As all the norms on freedom came from the pre-TU period, they were all approved without any need for debate[15].

Although all these liberal principles were never directly discussed during the conferences, proposals to introduce urgent telegrams in the period 1865-75 and the ensuing debates brought to the surface different attitudes to the freedom of correspondence. The supporters of urgent telegrams desired to furnish an extra service and thus increase freedom of action[16], while those against them spoke in terms of a possible inequality and injustice, which they felt undermined individual freedom.

In Paris, the Italian Minotto[17], backed by the Belgian Fassiaux[18], introduced the issue by speaking of his sorrow over the lack of norms covering urgent telegrams. The two received point-blank opposition, first from the French Jagerschmidt, who deemed the urgent telegram a

---

internationale, soit à des objets d'intérêt public déterminés de concert par lesdites administrations», ibid.

13 «La transmission des dépêches a lieu dans l'ordre suivant: 1) Dépêches d'État ; 2) Dépêches de service ; 3) Dépêches privées. […] Les dépêches de même rang sont transmises par le bureau de départ dans l'ordre de leur réception. Entre deux bureaux en relation directe, les dépêches de même rang son transmises dans l'ordre alternatif», art. 11, *Convention télégraphique internationale de Paris 1865*.

14 A brief debate on the order of sending telegrams appeared the *Journal Télégraphique*: P. G. H. Linckens, "Quelques considérations, relativement à l'ordre de transmission des dépêches," *Journal Télégraphique* 3 (1870) : 29-31 ; Francesco del Guerra, "Sur l'ordre de la transmission des dépêches," *Journal Télégraphique* 4 (1870) : 60-61.

15 The only exception, the very short debate on the nature of state telegrams and their effect on the order of sending telegrams at the Vienna Conference. *Documents de la Conférence Télégraphique Internationale de Vienne*, 219-224.

16 On the arguments of those in favour of urgent telegrams see : J. Vinchent, "De l'ordre de la transmission des dèpêches. – Résponse aux observations de M. del Guerra", *Journal Télégraphique* 6 (1870): 71-73.

17 «M. le Chevalier de Minotto exprime le regret que la Convention nouvelle n'admette pas les dépêches privées dites *urgentes* qui, dans certains pays, jouissent de la priorité sur les autres, moyennant une taxe plus forte», *Documents Diplomatiques de la Conférence Télégraphique Internationale de Paris*, 128.

18 «M. Fassiaux partage le même regret: il pense que ces dépêches répondent à un besoin de notre époque, et il cite des cas où elles lui paraissent avoir une utilité réelle», ibid.

"regrettable privilege"[19] and then Curchod, who stressed that all private telegrams were to be considered urgent without any distinctions and that the creation of a special category would be taking a step backwards to pre-TU days[20]. Three years later D'Amico spoke for Italy and again proposed the introduction of urgent private telegrams, arguing that a high surcharge would keep down the number sent[21]. Yet again Jagerschmidt gave a harsh reply and answered that the urgent telegram was a clear violation of individual freedom, quoting part of the report of the parliamentary committee that had examined the question in France:

> With urgent telegrams you will end by creating a monopoly service in the hands of the rich. By means of urgent telegrams you will disorganise the transmission service. You will remove all security from ordinary telegrams. It will be impossible to ensure that an ordinary telegram accepted at 10 o'clock in the morning will be transmitted at six in the evening. If you adopt this plan you will throw the service into disrepute. In the eyes of the public you will be removing all security and guarantees from the service[22].

He then added:

---

[19] Ibid.

[20] «M. Curchod [...]fait remarquer que toute dépêche télégraphique doit arriver promptement, qu'il faut éviter d'entraver la marche d'une catégorie de dépêches au profit d'une autre catégorie, et que la mesure proposée serait un pas en arrière», ibid.

[21] «M. D'Amico motive cet amendement sur l'utilité, incontestable dans certains cas, de donner aux dépêches ayant un caractère d'urgence particulier la priorité sur les autres dépêches privées. Ce droit existe en Belgique et en Italie, et il a été exercé jusqu'à ce jour sans réclamation. Il en demande donc l'introduction dans le service international. Mais, pour éviter que les dépêches urgentes ne se généralisent et ne nuisent aux autres, il les frapperait d'une taxe élevée, une taxe triple de la taxe ordinaire», *Documents de la Conférence Télégraphique Internationale de Vienne*, 242.

[22] «Avec la dépêche urgente, vous arriveriez à constituer le télégraphe à l'état de monopole entre les mains des gens riches. Par la dépêche urgente, vous désorganiseriez le service des transmissions ; vous enlèveriez à la dépêche ordinaire toute sécurité ; il serait impossible d'assurer qu'une dépêche ordinaire déposée à 10 heures du matin serait transmise à 6 heures du soir. Si vous adoptiez cette combinaison, vous jetteriez la défaveur sur le service télégraphique ; aux yeux du public, vous enlèveriez à ce service toute sécurité et toute garantie», ibid., p. 243.

> There are circumstances in which the right would constitute a real service; but these cases are exceptional, and isolated interests cannot let us lose sight of the danger of seeing the big commercial and industrial concerns monopolising the telegraph for their profit. All telegrams are urgent, and setting up categories would modify the liberal nature of the service[23].

Faced by these accusations of illiberalism and inefficiency, Vinchent stepped in to defend the Italian proposal by saying that urgent telegrams had been used in Belgium for the previous 13 years without any problems at all[24]. It is interesting to note how Vinchent and D'Amico defended their positions by referring to national experiences[25], while Jagerschmidt echoed Curchod's speech in Paris and appealed to custom. On two different occasions both of them had justified their opposition to urgent telegrams by arguing they had been abolished in a past conference.[26] In this case, unsurprisingly, the conservative tendency of the delegates once again re-enforced the passive acceptance of a familiar norm.

Yet again at the Rome Conference, the Italian delegation proposed re-introducing urgent telegrams. D'Amico, as chair, introduced the amendment with greater force than in the past. First of all, he stressed that

---

[23]«Il est des circonstances où le droit constituerait un véritable bienfait ; mais ces cas sont exceptionnels, et des intérêts isolés ne sauraient faire perdre de vue le danger de voir les grands établissements commerciaux et industriels monopoliser le télégraphe à leur profit. Toute dépêche télégraphique est d'ailleurs urgente, et l'établissement de catégories modifierait gravement le caractère libéral du service télégraphique», ibid.

[24] «Quant aux inconvénients de la dépêche urgente, ils ne se sont pas manifestés en Belgique où cette dépêche est autorisée depuis 13 ans. Le nombre des dépêches privées transmises par priorité est suffisant pour prouver que ce mode de correspondance est apprécié du public, et assez restreint pour ne causer aucun dommage au service ordinaire. Les résultats sont tels que l'office belge n'est pas éloigné de réduire le prix de la transmission par priorité. La dépêche urgente pourrait donc entre dans le service international, et M. Vinchent est profondément convaincu qu'un avenir peu éloigné justifiera cette opinion. Les réductions de taxes ont pour effet de multiplier les correspondances et d'encombrer les lignes. Il suffit de quelques pas nouveaux dans la voie de la diminution des tarifs pour que la dépêche urgente devienne une nécessité», ibid. p. 244.

[25] «M. D'Amico constate que, depuis 6 ans que ce mode de correspondance est établi en Italie, il n'a donné lieu à aucune plainte, ni à aucun abus», ibid.

[26] «M. Jagerschmidt rappelle que la question a été agitée et résolue négativement en 1865 aux Conférences de Paris», ibid., 243 ; «M. Curchod rappelle que cette disposition a dû être supprimée par la Convention de Berne, et que cette suppression a été le fruit de l'expérience», *Documents Diplomatiques de la Conférence Télégraphique Internationale de Paris*, 128.

there was a specific need for some clients, who every so often required the quickest service possible[27]. He then spoke of a decade-long positive experience both in a relatively large country like Italy and a much smaller one like Belgium, but with an intense activity in telegrams[28]. Finally, aware of the main criticism the opposition would make, he argued that an urgent telegram was not so much a privilege as a diversification in the service, just like exclusive express trains for first class passengers, which had never been the cause of protests about illiberality[29]. Vinchent too made a more emphatic communication in favour of the Italian proposal and spoke against the presumed illiberality of urgent telegrams:

> The urgent telegram, M. Vinchent added, is a provision that can make the service precious for all, without doing any appreciable wrong to anyone, and each delegate, of course will give it just value, if he has personally used it.[30]

Though the arguments of D'Amico and Vinchent had partly convinced the French delegation, liberal opposition to the amendment continued to be very strong. The Russian delegate Lüders found it unjust, holding that an

---

[27] «Dans beaucoup de circonstances, le télégraphe n'est employé que pour gagner du temps sur la poste. Mais quelquefois le public éprouve impérieusement le besoin d'une transmission immédiate et, pour l'obtenir, il serait disposé à faire tous les sacrifices pécuniaires qu'on lui demanderait. C'est à cette exigence que l'amendement italien a pour but de donner satisfaction», *Documents de la Conférence Télégraphique Internationale de Rome*, 279.

[28] «Cette mesure a donné de bons résultats en Belgique que est également favorable en Italie, c'est-à-dire, dans un grand État, mais ou le mouvement des correspondances est moins actif qu'en Belgique. Elle a donc reçus de la pratique et dans des conditions différentes une double consécration. Quant à l'objection que ces dépêches constitueraient un privilège, M. le Président ne la croit pas fondée le même raison contre beaucoup d'autres mesures qui n'ont jamais donné lieu ; car on pourrait invoquer», ibid.

[29] «Quant à l'objection que ces dépêches constitueraient un privilège, M. le Président ne la croit pas fondée ; car on pourrait invoquer la même raison contre beaucoup d'autres mesures qui n'ont jamais donné lieu à aucune opposition, contre les trains express, par exemple, accessible seulement aux voyageurs de première classe. L'expérience prouve, d'ailleurs, que le privilège des dépêches urgentes ne prend pas une telle étendue qu'il devienne une entrave pour les correspondances générales», ibid., 279.

[30] «La dépêche urgente, ajoute M. Vinchent, est une disposition qui peut rendre des services précieux à tous, sans faire à personne de tort appréciable, et chacun des délégués, certainement, l'estimerait à sa juste valeur, s'il avait personnellement à en faire usage», ibid.

urgent telegram advantaged the rich by worsening the service in general, while the British delegate Chambre was even more explicit:

> The speed of telegrams in the British service makes it useless and this privilege would disrupt the equality in treatment of which the English public is very jealous. A measure like this could lead, via the Press and in Parliament, to protests which the Post Office would be forced to heed.[31]

The debate in Rome ended with the inevitable rejection of the Italian amendment, though it had its last airing at St Petersburg. On this occasion D'Amico used the same arguments as in Rome, including the comparison to first-class express trains[32]. Thanks to the support of France and Germany, conditions were more favourable and the Austrian Brunner's opposition, based on a presumed difference between national and international telegraphs, was less adamant than in the past[33].

---

[31] «La rapidité des transmissions dans le service britannique la rend inutile et ce privilège romprait l'égalité de traitement dont le public anglais est très-jaloux. Une semblable mesure pourrait donner lieu, par la voie de la presse et dans le Parlement, à des protestations dont le Département des Postes serait obligé de tenir compte», ibid.

[32] «M. D'Amico fait observer que le public, de même qu'il recourt au télégraphe quand la poste ne lui offre pas une rapidité suffisante à, dans certaines circonstances, un intérêt considérable à ce que ses communications parviennent en quelque sorte plus vite que par le télégraphe. C'est à ces circonstances que prévoit le télégramme urgent. On a dit que son adoption constituerait un privilège. Mais, entendu de la sorte, le privilège serait dans l'ordre de toutes les choses humaines ; car c'est également un privilège que de jouir de la rapidité d'un train express qui est inaccessible aux voyageurs de la dernière classe ; c'est même un privilège que de se servir du télégraphe, dont la taxe si minime qu'elle soit, n'est pas à la portée de tous. Aux yeux de M. D'Amico, il n'y a pas de privilège, quand il s'agit d'une faculté accordée aux mêmes conditions à tous ceux qui la demandent ; d'une faculté dont le public des bourses modestes est lui-même heureux de profiter, le jour où il attache un intérêt de premier ordre à ce que sa dépêche arrive sans aucun délai. On a objecté, d'un autre côté, que l'adoption des télégrammes urgents serait une sorte de prime d'encouragement donnée au mauvais service. Cette objection ne parait pas sérieuse á M. D'Amico. [...] En terminant, M. le délégué de l'Italie ajoutera qu'au point de vue industriel, une administration trouve son avantage à rendre au public un service réel qui lui apporte, à elle, une rémunération légitime», *Documents de la Conférence Télégraphique Internationale de St Petersburg*, 310.

[33] «C'est que le résultat favorable constaté par plusieurs administrations pour leur service intérieure, la modicité des taxes fait que beaucoup de communications n'ont qu'une importance secondaire; pour la correspondance internationale, dont le prix est sensiblement plus élevé, toutes les dépêches sont importantes, toutes son des dépêches urgentes et le public ne comprendrait pas que l'on fit une distinction

Yet again the Swiss and British delegations manifested their opposition and when the assembly voted its approval to urgent telegrams, the two announced that they could use their veto, given that their governments' mandates were against the disposition[34]. Put bluntly, Switzerland and Great Britain would not have allowed urgent international telegrams to transit over their territory. The impasse was partially resolved with the introduction of an optional priority service[35]. So the question of an urgent telegram service was resolved with a stratagem conceding individual administrations freedom of decision. What the debate also showed was that many, often contrasting, interpretations existed around the concept of freedom and telegraphs.

4.1.2 Hours of service, languages and codes

The fulcrum of long distance international communication was the definition of a reciprocally recognised protocol, which implied not only the use of the same basic code but also the rigorous synchronization of time. The conferences consequently forged conventions and rules and

---

entre elles. M. Brunner douterait donc le dernier cas le résultat fût aussi favorable que dans le premier et il craindrait, au contraire, qu'il ne se produisît une tendance à se contenter un service médiocre», ibid., 311.

[34] Regarding Great Britain : «M. Chambre tient à constater qu'a Rome il a déclaré que ses instructions ne lui permettaient pas d'accepter la dépêche urgente. Ses instructions sont les mêmes aujourd'hui», ibid., 312. Regarding Switzerland : «M. Hammer doit déclarer que les instructions du Gouvernement fédéral lui enjoignent de s'opposer à l'adoption des dépêches urgentes. Ces instructions s'appuient sur ce motif qu'une exception faite en faveur de la richesse présenterait quelque chose d'odieux dans un pays qui a un profond sentiment de l'égalité. D'une autre côté, il voudrait savoir quelle garantie il y aurait pour que les dépêches ordinaires fussent transmises dans un temps déterminé, quand leur transmission viendrait à être primée par des correspondances privilégiées. Si l'on pouvait prendre pour règle de l'ordre de la transmission, l'importance du contenu des dépêches, au lieu d'accorder la priorité à l'expéditeur qui peut payer une plus forte taxe, l'urgence se justifierait logiquement. Mais une pareille appréciation est impossible, et dans l'application, la priorité se réduit à une question d'argent. Or, la chose présente, dès lors, ce double danger, d'une part, que le public y voie une mesure fiscale de la part des administrations pour retirer un plus grand profit sans accroissement de travail et, de l'autre, que le personnel télégraphique ne soit enclin à négliger la correspondance ordinaire pour recommander l'urgence aux expéditeurs. Ces inconvénients seraient d'autant plus sensibles que le nombre des dépêches urgentes s'augmenterait et si, au contraire, le nombre en était insignifiant, la mesure ne présenterait ni l'intérêt ni l'utilité qu'y trouvent ses partisans», ibid., 313.

[35] Ibid., 487.

established the code the operators would adopt, the languages the clients were to use, office opening and closing hours and the criterion for establishing time in all the offices.

The Morse telegraph and code had already been chosen in the pre-Paris conferences, and were therefore never called into doubt. Where there was much debate was over the languages allowed in international telegrams. In Paris it was decided that any of the languages spoken in member states could be used, and that in deference to national sovereignty, each state could choose which languages were viable on its territory[36]. Nevertheless, in order to avoid transmission errors, Minotto had tried to establish a uniform language. He started by doubting if it was legitimate to ask for reimbursement for mistakes made by operators in dealing with telegrams in languages they did not know[37], a doubt that was quashed by the assembly. And then he went on to suggest using French, in that it was the diplomatic language for service telegrams between different administrations[38], an idea that was accepted[39]. Both motions initiated long reflections over the problem. At Vienna in fact numerous amendments were presented about the languages to use internationally. The most debated proposal was that of the Portuguese delegate Falcoïano, which foresaw the adoption of one universal language to be chosen by the administrations, leaving all the telegrams worded in different languages to pay extra charges as coded messages[40]. Though the delegates found the Portuguese proposal

---

[36] «Toute dépêche peut être rédigée en l'une quelconque des langues usitées sur le territoire des États contractants. Chaque État reste libre de désigner, parmi les langues usitées sur son territoire, celles qu'il considère comme propres à la correspondance télégraphique.», art. 9, *Convention télégraphique internationale de Paris 1865*.

[37] «Sur le paragraphe premier, M. le Chevalier de Minotto pense que si l'on oblige les Administrations télégraphiques à recevoir des dépêches dans l'une quelconque des langues en usage sur le territoire des États contractants, il conviendrait de les exonérer de tous remboursements pour erreurs commises sur le territoire où l'erreur s'est produite», *Documents Diplomatiques de la Conférence Télégraphique Internationale de Paris*, 123.

[38] «M. le Chevalier de Minotto présente une autre proposition ainsi conçue : "Les dépêches de service devront toujours être rédigées en français ou dans la langue du pays auquel elles sont adressées"», ibid.

[39] «En règle générale, les dépêches de service sont rédigées en français ; toutefois, les diverses administrations peuvent s'entendre entre elles pour l'usage d'une autre langue», art. VI du *Règlement de service international destiné à compléter les dispositions de la Convention Télégraphique de Paris 1865*.

[40] «Tant qu'il ne sera pas créé une langue universelle pour la communication télégraphique, la langue qui sera choisie de préférence à toutes les autres sera

interesting, they were quick to identify its main problem, in that those who did not know the universally shared languages would be forced to pay higher charges[41]. Given the defects in all the proposed amendments, Vinchent and Brunner suggested keeping the Paris norm, which was then confirmed in Vienna[42] with the surprise introduction of Latin, too[43]. Three years later the language question came up again in Rome, the debate springing this time from two different points of view. Some held the need for one or more conventional languages while others desired the maximum number of languages possible. Among those in the former group, Portugal held that each state would have to word telegrams in its own language[44] so as to avoid mistakes while Ludërs and Brandstrom, the delegates for Russia and Sweden respectively, suggested using the three most widely known languages (English, French and German)[45]. In the latter group, the

---

provisoirement et exclusivement adoptée. Toute dépêche écrite dans une langue autre que celle qui a été choisie sera considérée comme dépêche chiffrée», *Documents de la Conférence Télégraphique Internationale de Vienne*, 230.

41 «M. D'Amico combat, au point de vue des relations commerciales, l'amendement du Portugal, qui se traduirait en fait par une augmentation de taxe pour toutes les personnes qui ne connaîtraient pas la langue choisie», ibid., 230-231.

42 «M. Vinchent, examinant les trois systèmes entre lesquels la Conférence doit se prononcer, voit des inconvénients à tous ; mais l'état actuel des choses est encore celui qui en présente le moins. Il en demande donc le maintien. M. Brunner le recommande également à la Conférence comme déjà sanctionné par l'expérience, et il croit que les deux premiers alinéas, entendus comme l'a expliqué M. Vinchent, doivent être conservés», ibid., 232-233.

43 «Toute dépêche peut être rédigée en l'une quelconque des langues usitées sur les territoires des États contractants, et en langue latine», art. 9, *Convention télégraphique internationale de Vienne 1868*.

44 «Le Gouvernement portugais propose de n'admettre que les langues désignées comme propres a la correspondance télégraphique par l'État auquel appartient le bureau d'origine et de considérer comme secrète toute dépêche écrite dans une autre langue», *Documents de la Conférence Télégraphique Internationale de Rome*, 264.

45 «Le Gouvernement russe reconnaîtrait, en principe, comme propres à la correspondance internationale les langues française, allemande et anglaise, en laissant à chaque État la faculté de désigner laquelle ou les quelles de ces trois langues usitées sur les territoire des États contractants ne seraient pas refusées ; mais les bureaux ne pourraient être rendus responsable des erreurs commises dans leur transmission. Les États limitrophes auraient, d'ailleurs, le droit de s'entendre entre eux pour l'usage d'une autre langue. [...] M. Brändström appuie la proposition de la Russie, mais il voudrait aller plus loin. Les employés ne peuvent connaître toutes les langues ; il faut donc limiter le nombre de celles qui sont

Austrian and Turkish delegations, whose states were melting pots of numerous ethnic groups and tongues, upheld instead that telegrams should be admitted in all existing languages and not only those approved by the governments concerned[46]. In the end, the standing norm was judged the most appropriate and all amendments were rejected. Unsurprisingly, the principle of state sovereignty again prevailed and once again, in spite of an intense debate, the delegates remained loyal to a Paris-approved norm.

A close link existed between the topics of the language adopted in telegrams and encrypted codes. The Paris Convention opened the possibility of transmitting government, service and private telegrams in secret codes, where and when allowed[47]. However the last section of article 9 clearly forbid the formulation of conventional codes[48], which were practices adopted in the business world to save money. The difference between a secret and a conventional code was that the latter was based on a known system of decoding, which the telegrapher could possibly have access to. The question of conventional code telegrams emerged quite forcefully at the Vienna Conference, where many delegates commented that it was not always easy to tell the difference between them and those couched in ordinary language. In fact, given there was no normative difference between conventional and secret codes, many in business circles avoided surcharges on secret codes by resorting to languages composed of ordinary words they gave different meanings to[49]. Vinchent wanted to resolve the question by continuing to treat the two kinds of codes on the same level.

---

reconnues propres au service international aux langues les plus répandues, l'anglais, l'allemand et le français», ibid.

[46] «Avant que la question de la limitation des langues soit soumise à la votation, Izzet Effendi rappelle que son Gouvernement a fait également une proposition qui concerne l'emploi des différentes langues. Yanco Effendi ajoute que cet amendement a pour objet d'admettre toutes les langues usitées, pourvu qu'elles soient écrites en caractères latins, et c'est pour ce motif que son Gouvernement propose la suppression du paragraphe 2 de l'article 9», ibid., 267.

[47] «Les dépêches d'État et de service peuvent être composées en chiffres ou en lettres secrètes, soit en totalité, soit en partie. Les dépêches privées peuvent aussi être composées en chiffres ou en lettres secrètes, lorsqu'elles sont échangées entre deux États contractants qui admettent ce mode de correspondance», art. 9, *Convention télégraphique internationale de Paris 1865.*

[48] «Les dépêches en langage ordinaire ne peuvent contenir ni combinaisons de mots, constructions, ni abréviations inusitées», ibid.

[49] «M. Radojkovits signale la difficulté, pour les bureaux télégraphiques, de distinguer dans la plupart des cas les dépêches inintelligibles des dépêches ordinaires. Chaque profession a un langage qui lui est propre, et dont l'emploi pourra être considéré comme abusif par un agent qui ne sera pas au courant de ce

Many of the delegates were not of the same idea[50], and Curchod started to defend business interests by proposing the abolition of the last paragraph of article 9[51]. In this way telegrams in conventional languages would be treated as ordinary telegrams and avoid paying the secret code surcharge[52].

---

langage. Le public se trouvera donc livré à l'arbitraire des bureaux télégraphiques, et les difficultés qu'on lui suscitera n'auront d'autres résultats que de l'irriter et de le rendre plus ingénieux dans le choix des moyens d'échapper aux restrictions établies contre lui», *Documents de la Conférence Télégraphique Internationale de Vienne*, 237.

[50] «M. le Colonel Goldsmid verrait avec regret une disposition qui assimilerait les dépêches commerciales aux dépêches secrètes. [...] M. Brandstrom serait plutôt porté à supprimer le dernier paragraphe de l'article qui, dans les pays parlant des langue d'origine germanique, n'est pas susceptible d'application. S'il n'en fait pas la proposition, c'est qu'il craindrait de paraître encourager le mode de correspondance interdit par ce paragraphe et l'emploi des constructions ou combinaisons inusitées. [...] M. le Colonel de Chauvin croit que la diversité des opérations commerciales rend impossible de donner une clef applicable à toutes les dépêches échangées entre deux correspondants. [...] M. Brunner ne le conteste pas; mais le même fait peut avoir lieu dans le service ordinaire. Ce qu'on cherche, c'est le moyen de traiter les dépêches commerciales comme dépêches ordinaires ; on peut le faire, quand il est possible de leur attribuer un sens au moins apparent. [...] M. Jagerschmidt rétablit la pensée de l'amendement, qu'on semble perdre de vue : il ne s'agit pas d'imposer au public l'obligation d'être clair dans ses dépêches, mais seulement de les frapper d'une surtaxe quand elles donnent lieu à un travail supplémentaire. D'un autre côté, la Conférence désire que cette surtaxe ne s'applique pas aux dépêches de commerce», ibid., 238-240.

[51] «Les difficultés de la question amènent M. Curchod à faire une proposition qui a l'avantage d'être précise et facile à exécuter. Elle consisterait à effacer le paragraphe 9 et à supprimer ou du moins à simplifier beaucoup le collationnement d'office qui, dans la pratique, n'a pas l'efficacité qu'on lui suppose. Le collationnement serait maintenu comme opération accessoire donnant lieu à une surtaxe. Les expéditeurs ne seraient pas fondés à se plaindre des irrégularités commises dans la transmission d'une dépêche inintelligible, puisqu'il dépendrait d'eux de prendre des garanties spéciales ; de leur côté, les offices n'auraient plus à faire entre les dépêches des distinctions toujours fort délicates. M. Curchod demande donc que le dernier paragraphe de l'article 9 soit supprimé, sauf à introduire au règlement les modifications nécessaires», ibid., 240.

[52] «M. Vinchent met en relief la gravité de cette proposition, qui aura pour effet d'autoriser la circulation, comme dépêches ordinaires, de dépêches réellement secrètes, consistant en assemblage de mots qui n'auront aucun sens», ibid.

Defeated at Vienna but on the attack again at Rome, Vinchent proposed an amendment which put the two codes on the same level[53]. This time he came up against the private companies, which habitually used conventional codes on their submarine cables[54]. In addition, Curchod, now representing both the French Atlantic cable company and Switzerland, came in on the attack[55], so forcing Vinchent to tone down his amendment to the point of considering conventional codes like ordinary language[56], just as had happened at Vienna. Allowing conventional codes was an evident concession for the business world, the biggest group of users beside the Press.

---

53 «M. Vinchent croit utile de faire remarquer que son amendement contient autre chose qu'une modification de forme. C'est avec l'usage, mais non avec les dispositions actuelles qu'il est d'accord. Ainsi, il précise que les dépêches en langage convenu sont des dépêches secrètes, ce qui est une innovation», *Documents de la Conférence Télégraphique Internationale de Rome*, 270.

54 «M. Despecher reconnait avec M. Vinchent qu'au point de vue de la définition, les correspondances dont il s'agit doivent être considérées comme secrètes. Mais le nombre de ces dépêches entre pour une telle proportion dans le trafic des câbles sous-marins que les Compagnies ne pourraient réagir contre les habitudes du public, en les soumettant désormais à la double taxe. Il se réserve donc de demander, au moment voulu, qu'elles continuent à être traitées comme les dépêches ordinaires.», ibid., 272.

55 «En ce qui concerne maintenant la rédaction proposée par la Belgique pour l'article 9, M. Curchod fait observer qu'elle lui paraît préjuger déjà les décisions à prendre ultérieurement sur les conditions faites aux dépêches en langage conventionnel, puisqu'elle stipule qu'elle sont considérées et traitées comme dépêches secrètes. Ces dépêches en langage conventionnel ne sont point, d'ailleurs, à proprement parler, des dépêches dont l'expéditeur cherche à dissimuler le sens ; car, le plus souvent, il est tout disposé à en donner la clef aux bureaux télégraphiques. Ce sont des dépêches commerciales, auxquelles le public donne cette forme pour en restreindre le nombre de mos. Or, la Conférence de Vienne a voulu accorder une certaine tolérance aux dépêches commerciales, et cette tolérance est consacrée par l'article V du Règlement. Dans les correspondances avec l'Amérique, les Compagnies ont été amenées, de la sorte, à distinguer trois catégories de dépêches : les dépêches en langage ordinaire dont le traitement est connu, les dépêches en langage conventionnel taxée comme les premières, mais pour les quelles les Compagnies ne répondent pas des erreurs de transmission et les dépêches en chiffre, soumises à la double tax», ibid., 273-274.

56 «Pour donner satisfaction aux observations de M. Curchod, M. Vinchent est disposé à modifier la rédaction de son amendement de la manière suivante. Dans le troisième paragraphe de l'article 9, il supprimerait les mots "et traitées" et dans l'alinéa 3 de ce même paragraphe, ajouterait les mots "incompréhensibles pour les offices en correspondance" aux mots "les dépêches contenant des passages en langage convenu"», ibid.

A last important element of standardization concerned the synchronization and coordination of the offices involved in the international service[57]. For by then the TU was monitoring a telegraph network that spread from Great Britain and Spain in the west to British India and Asian Russia in the east.

Consequently, right from the beginning a standard time had to be fixed in each office independently from the latitude concerned. Secondly, it had to be posted when offices in other countries were open and therefore able to receive telegrams. The problem obviously did not arise in the most important offices, which were open 24/24, but in those operating only in daytime. So article 2 of the Paris Convention established that daytime offices had to be open from seven in the morning to nine in the evening in the summer semester and with an hour's later start in winter time[58]. In spite of Brunner's opposition[59], it was decided that all offices in a country had to work with the same time as the capital[60].

While office opening/closing hours remained the same until St Petersburg, some debating took place in Rome about the way in which the time had to be set within each office. The Turkish delegate suggested synchronizing the time in all offices to that of a single meridian while Meydam proposed using local time in each office[61]. Vinchent instead

---

[57] On the relation between telecommunications and the synchronization of time see: John B. Thompson, *The Media and Modernity. A social theory of the media* (Stanford: Stanford University Press, 1995); Stefano Bregni, "A Historical Perspective on Telecommunications Network Synchronization," *IEEE Communications Magazine* 36 (1998): 158-166.

[58] «Entre les villes importantes des États contractants, le service est autant que possible permanent, le jour et la nuit, sans aucune interruption. Les bureaux ordinaires, à service de jour complet, sont ouvert au public : Du 1er avril au 30 septembre, de 7 heures du matin à 9 heures du soir, Du 1er octobre au 31 mars, de 8 heures du matin à 9 heures du soir. Les heures d'ouverture des bureaux à service limité sont fixées par les administrations respectives des États contractants», art. 2, *Convention Télégraphique Internationale de Paris 1865*.

[59] «"L'heure de tous les bureaux d'un même État est celle du temps moyen de la capitale de cet État", une discussion s'engage entre M. Brunner et plusieurs membres de la Commissions. M. Brunner propose la suppression de ce paragraphe ; il déclare que cette disposition n'a jamais pu être appliquée en Autriche, et il exprime la pensée qu'il doit être de même en Russie», *Documents Diplomatiques de la Conférence Télégraphique Internationale de Paris*, 116-117.

[60] «L'heure de tous les bureaux d'un même État est celle du temps moyen de la capitale de cet État», ibid.

[61] «M. le Président met d'abord en délibération l'amendement du Gouvernement ottoman relatif à la désignation d'une méridien unique pour la fixation des heures employées par les bureaux télégraphiques. Yanco Effendi fait observer que le public a intérêt à savoir exactement la durée de la transmission des dépêches qu'il

spoke in favour of keeping the same time within a state, with its capital as benchmark, justifying the measure with the need to coordinate the telegraph service with national transport systems and in particular the railways[62]. Thus the criterion established at Paris was conserved.

Midway between normative and technical standardization lay the norms concerning sending telegrams to places outside the network. Though apparently of secondary importance, discussing them took up two sittings at the Paris Conference[63]. Some delegates like Vinchent were in favour of sending telegrams on with the fastest means available, and therefore advised inserting in the norm compulsory registration or express service[64]. However, while this solution was immaterial to well-organized administrations working in a small territory like Belgium, it would have meant a radical hike in costs for countries like the Russia, where a large part of its enormous territory was outside the network. The contributions of Curchod[65] and Brunner[66] followed by that of the chair himself[67],

---

reçoit ; ce qu'il ne peut connaître, avec des méridiens différents, qu'au moyen d'un calcul dont il ne possède pas toujours les éléments. […] M. Meydam voit des inconvénients à employer le méridien de la Capitale dans les États très-étendus. Pour se rapprocher davantage des usages, il propose d'adopter pour chaque bureau le temps moyen de la localité qu'il dessert», *Documents de la Conférence Télégraphique Internationale de Rome*, 248-249.

[62] «M. Vinchent ne pourrait pas accepter le proposition de M. Meydam, qui entraînerait pour la Belgique l'obligation de remanier complètement les horaires suivis pas les services similaires des Chemins de fer et des Poste. Afin de tenir compte des exigences particulières aux grands pays, il propose que la rédaction soit modifiée, de façon que chaque Gouvernement ait la faculté de désigner une heure unique pour tout l'État ou pour chaque grand région territoriale», ibid.

[63] *Documents Diplomatiques de la Conférence Télégraphique Internationale de Paris*, 133-148.

[64] «M. Vinchent voudrait arriver à effacer du paragraphe la distinction qu'il établit entre l'envoi par lettre ordinaire et l'envoi par lettre recommandée. Il propose, dans ce but, la gratuité du transport par lettre recommandée dans les limites du même État. [...] Sur le même paragraphe, M. Vinchent exprime le regret de ne trouver Dans le Projet aucune disposition relative aux exprès», ibid.,135-136.

[65] «M. Curchod intervient alors pour appuyer l'insertion, dans l'article, d'une mention relative aux exprès. Il comprend qu'il soit impossible de prendre à cet égard un engagement collectif ; mais il voudrait que pour éviter une confusion qui se produit déjà aujourd'hui, et qui lui paraît devoir s'augmenter encore dans le système du Projet, il y eût pour les exprès une seule règle, que chaque État fût libre de suivre ou de rejeter, mais qui serait appliquée partout où le service serait maintenu. M. Curchod propose en conséquence de dire, dans le paragraphe, que les dépêches seraient adressés par la poste « ou par exprès», et de déterminer, dans le Règlement, les conditions uniformes du service des exprès», ibid., 137.

reformulated the whole of the norm and thus managed to resolve the question. It left each state the discretion to send the telegram in the way that was most suitable for the needs of its administration, with in mind the common objective, to get the telegram to destination as quickly as possible[68].

Three years later at Vienna the question of sending telegrams on by post arose again during the discussion of article 42 (then 46)[69], about the surcharge to apply in exceptional cases. The Paris Convention had provided for a small sum to be added in the case of this extra service. However at Vienna the Spanish delegate, backed by Curchod, spoke for the abolition of the surcharge and therefore *de facto* the introduction of no

---

[66] «Sur cette observation, M. Brunner demande la parole. Il admet, quant à lui, l'utilité d'un service de transport plus rapide que la poste, soit par exprès, soit par tout autre moyen ; mais il admet aussi, avant tout, l'impossibilité de l'imposer à certains États. Il croit en conséquence utile de l'inscrire dans l'article, sans en faire une obligation, et sans en tracer les règles, ni dans la Convention ni dans le Règlement. Il propose d'insérer après les mots : « l'envoi a lieu par la poste, par lettre simple ou recommandée», la phrase suivante : «ou par tout autre moyen que l'Administration du bureau destinataire juge convenable d'adopter». Ces moyens seraient simplement communiqués aux divers États par celui qui les aurait introduits dans son service, et mis pas eux à la disposition du public», ibid., 137-138.

[67] «Avant d'ouvrir la discussion sur la suite des articles du Projet, M. le Président rappelle que la Commission a demandé une rédaction nouvelle pour l'article 16», ibid., 143.

[68] «Chacun des États contractants se réserve d'organiser, autant que possible, pour les localités non desservies par le télégraphe, un service de transport plus rapide que la poste ; et chaque État s'engage envers les autres à mettre tout expéditeur en mesure de profiter, pour sa correspondance, des dispositions prises et notifiées, à cet égard, par l'un quelconque des autres États», art. 17, *Convention Télégraphique Internationale de Paris 1865*.

[69] «Les dépêches recommandées ou avec accusé de réception, à envoyer par la poste ou à déposer poste restante, sont affranchies, comme lettres chargées, par le bureau télégraphique d'arrivée. Le bureau d'origine perçoit les taxes supplémentaires suivantes : un demi-franc par dépêche à déposer poste restante dans la localité desservie, ou à envoyer par la poste, dans les limites de l'État qui fait l'expédition ; Un franc par dépêche à envoyer en Europe hors de ces limites, sur le territoire des États contractants ; Deux francs et demi par dépêche à envoyer au-delà. Les dépêches non recommandées sont acquittées, s'il y a lieu, par le destinataire, aucune taxe supplémentaire n'étant perçue par le bureau d'origine», art. 46, *Convention Télégraphique International de Vienne 1868*.

extra charge for telegrams sent on by post. The debate was intense, but at the end the article was, unsurprisingly, left as it was[70].

4.2 Standardizing technology

Two characteristics regarding the technological standardization of telegraphs up to the 1850s need to be taken into consideration: 1) there was an evident convergence on transmission apparatus and material for building the lines; 2) on an international level technical standardization could only concern the wires and office apparatus.

4.2.1 Convergence

Land and submarine telegraphs were technologically very different. As the term indicates, land telegraphy, which included most of the world's lines, covered the inexpensive and technologically very simple infrastructures constructed on land. Submarine cables were instead technologically complex and costly, as were the techniques for transporting and laying them. This section will consider only landlines because the privately-run submarine cables never underwent any process of standardization under the auspices of the TU. Things were different in the case of landlines, where the technologies adopted by the various administrations had tended to converge in pre-TU days.

The land telegraph networks worked through offices linked to one another via the lines connecting them. The telegraph instrument, installed in the post office building, brought about the transmission of messages via electric impulses. At the beginning of the 1850s there were at least three different types: 1) the electromagnetic, type Morse; 2) the needle; 3) the dial.

The Morse telegraph was based on a physical principle discovered by François Arago, according to which electric power passed through a metal wire twisted around a piece of iron is magnetized[71]. More simply, when electric power was applied in the Morse apparatus, the previously magnetized piece of iron attracted another in the shape of a pointed hammer. A paper ribbon unwound between the two metal pieces by means

---

[70] *Documents de la Conférence Télégraphique Internationale de Vienne*, 333-338.

[71] Dominique François Jean Arago, "Note concernant les Phénomènes magnétiques auxquels le mouvement donne naissance," *Annales de chimie et de physique* 32 (1826): 213-223; James Lequeux, *François Arago, un savant généreux* (Paris: EDP- Sciences, 2008).

of a clockwork mechanism and left a mark imprinted on it. According to the length of the electric impulse, this mark could be a dot or a dash. The length of the impulse was established by the transmitting telegraph operator pressing a lever which guaranteed the opening of the electric circuit, and therefore the release of power to another apparatus. The messages were therefore sent and received in a dot-dash code, which remained imprinted in relief on the paper. There was a modified version of Morse which made it possible to print the signs in ink, thus making reading easier for operators. In actual fact, the machine only had to ensure that when electric power was applied the hammer would imprint the mark, which according to the length of the electric impulse, could be a dash or a dot. This made it an extremely simple mechanism and a fairly easy one to repair, even by the operators themselves[72].

The so-called needle telegraphs, like the model patented by Wheatstone and Cooke in 1837, were based instead on another fundamental principle of electricity, discovered by Hans Christian Oersted, according to which if electric power is applied through the needle of a compass, the latter temporarily changes its magnetic pole[73]. Therefore, more simply, according to whether positive or negative electric power was applied, the needle on the Wheatstone telegraph dial moved to the right or the left. At the beginning, this telegraph had five needles, which were then reduced to two to facilitate use along the railroads. The code used was based on the combination of the two to five needles moving to the right and the left. This code was obviously more complex than Morse, and for this reason the latter soon came into use also on one-needle telegraphs, considering by rule the right and left hand oscillation respectively as a dot and as a dash. The Wheatstone telegraph still had a significant drawback. The messages could not remain imprinted on paper, and had to be interpreted and translated simultaneously by the operator, thus increasing possibilities of reception errors[74].

---

72 Shaffner, *The Telegraph Manual*, 422-456; Carlo Matteucci, *Manuale di telegrafia elettrica* (Torino: Unione Tipografica Editrice, 1861), 108-121.

73 Bern Dibner, *Oersted and the Discovery of Electromagnetism* (New York: Blaisdell, 1962); Ole Immanuel Franksen, H. C., *Oersted – A Man of Two Cultures* (Birlerod: Strandsberg Forlag, 1981); Roberto de Andrade Martins, "Resistance to the Discovery of Electromagnetism: Oersted and the Symmetry of the Magnetic Field," in *Volta and the History of Electricity* eds. Fabio Bevilaqua and Enrico Giannetto (Pavia/Milano: Universitá degli Studi di Pavia-Hoepli, 2003), 245-265.

74 Dionysus Lardner, *The Electric Telegraph Popularised* (London: Walton and Maberly, 1855) 177-192; Robert S. Culley, *A Handbook of Practical Telegraphy* (London: Longman, Green, Longman, Roberts, and Green, 1867), 168-175; Robert

There were also highly complex telegraphs, generally known as dial systems, including Breguet's[75]. In this type of mechanism the sender device was separate from the receiver, which already made it more complex than the other types. Messages were sent with a key which displayed all the letters of the alphabet around a circle, and to send messages the operator turned a knob on to the position of the required letter. According to the position of the letter on the key, the operator's gesture produced a given number of electric impulses which reached the receiver telegraph operator, making the needle on the dial move to the same letter. Due to the many inner gears, the Breguet telegraph was particularly complicated and therefore hard to repair. In general, however, dial telegraphs had the great advantage of making it possible to send and receive messages without employing conventional codes. Nonetheless, this advantage, which reduced errors occurring in translation from one code to another, was countered, as in needle telegraphs, by the impossibility of printing messages directly and therefore checking their correctness[76].

The advantages, also in economic terms, offered by the Morse system as compared to the others are therefore evident. Its simplicity made repairs easy, and they could often be carried out without having to call in the manufacturers. This meant repair and maintenance costs were vastly lower than those of more complicated telegraphs. The immediate printing of the messages on paper reduced the possibilities of error, and therefore stimulated greater utilization of the service[77]. It is scarcely surprising therefore that from the later fifties onwards the Morse telegraph was the main apparatus in use on international lines.

As we have already mentioned, telegraph lines were composed of three basic elements: 1) wires; 2) poles; 3) insulators[78]. The wires could be made of copper or iron - metals with very different characteristics. Copper

---

Sabine, *The History and Progress of the Electric Telegraph* (London: Lockwood & C., 1872), 54-57.

[75] Jean Claude Montagné, *Histoire des Moyens de Télécommunication de l'Antiquité à la Seconde Guerre Mondial: du sifflet préhistorique à la télévision. Les hommes, les idées* (Bagneux: J. C. Montagné, 1995), 131-32.

[76] Charles Bontemps, *Les systèmes télégraphiques aériens – électriques, pneumatiques* (Paris : Dunod Éditeur, 1876), 73-80 ; Edward B. Bright, *The Electric Telegraph* (London : James Walton, 1867), 153-154.

[77] To compare the mechanisms in use at the time, see Matteucci, *Manuale di telegrafia elettrica,* 127-130.

[78] We consider here only the lines in use all over Europe. Underground cables had been placed in the main British and German cities, but they were expensive, not practical for international routes and therefore not taken into consideration by the TU.

makes a better electric wire than iron, but it is far less resistant to being twisted and subjected to strain. And outdoors, the wire was often subject to the effect of atmospheric agents like wind and snow, which could twist or strain it. Furthermore, to be fastened to the poles it needed to be pulled from one supporting pole to the next without suffering damage. It is easy to understand therefore why by the 1850s many states had already opted for iron[79].

The poles holding the wire lines overhead were obviously essential for the correct long-time working of the telegraph service. They were always made of wood, given that it is a very poor conductor, and thus guarantees a negligible dispersion of electricity. Another advantage of wood was that it was then the least expensive material. The type of wood used, however, could have heavy repercussions on maintenance costs and states tended to use the most readily available variety guaranteeing the longest life-span[80].

Insulators were another technological component essential to the development of telegraph lines. Their purpose was obviously to avoid power dispersion during communication by keeping the wire insulated from its iron support. Insulators were made of glass, ceramic or porcelain and were of all sorts of shapes, like bells, inverted mushrooms, etc. As for the materials, the least efficient was definitely glass, which was most likely to trap moisture and therefore disperse more electricity. The best was definitely porcelain, which was far more reliable in very damp locations[81].

So by the close of the fifties, five years or so before the TU was established, almost all the states in Europe were adopting electro-magnetic telegraphs Morse model, and building overhead telegraph lines with galvanized iron wires, wooden poles and ceramic insulators in different forms, according to geography and meteorology[82].

---

[79] Latimer Clark, *Birmingham wire gauge. A paper read at the British association for the advancement of science at Dundee, September 1867* (London: E. & Spon, 1867).

[80] Shaffner, *The Telegraph Manual*, 681 700.

[81] Ibid., 529-563.

[82] This information on the building materials and the above descriptions are mostly taken from an enquiry carried out by the *Bureau*: ITU-Corr., Dossier n. 79, 1869. The results of this survey were partially published in JT: "De l'établissement des lignes pour le service international (1ère article). Allemagne du Nord," *Journal télégraphique* 10 (1870) :137-141; "De l'établissement des lignes pour le service international (2ème article). Autriche. Bade. Bavière," *Journal télégraphique* 11 (1870) :153-157 ; "De l'établissement des lignes pour le service international (3ème article). Belgique," *Journal télégraphique* 12 (1870) : 169-175 ; "De l'établissement des lignes pour le service international (4ème article). Belgique

4.2.2 The role of the TU

The existence of a high level of convergence meant that technical standardization was not perceived as a priority when the Telegraph Union was being set up, though it probably was earlier, at the moment of the first bilateral and multilateral treatises. The technical standardization led by the Union was limited to a definition of the wires and the apparatus for transmission.

The 1865 Conference established in art. 3 the use of the Morse on international lines, with a preference for the ink printer model[83], while art. 1 advised, but did not impose, the use of a large diameter iron wire for direct lines between the big cities[84]. In reality, the convention project

---

(suite et fin)," *Journal télégraphique* 13 (1870):185-189 ; "De l'établissement des lignes pour le service international (5ème article). Danemark. France," *Journal télégraphique* 14 (1870) : 201-206 ; "De l'établissement des lignes pour le service international (6ème article). Grande Bretagne. Réseau métropolitain," *Journal télégraphique* 15 (1871): 217-220 ; "De l'établissement des lignes pour le service international (7ème article). Grande Bretagne. Réseau métropolitain (suite et fin)," *Journal télégraphique* 16 (1871): 233-237; "De l'établissement des lignes pour le service international (8ème article). Italie," *Journal télégraphique* 17 (1871): 249-253 ; "De l'établissement des lignes pour le service international (9ème article). Italie (suite et fin)," *Journal télégraphique* 18 (1871): 265-268; "De l'établissement des lignes pour le service international (10ème article). Norvège", *Journal télégraphique* 19 (1871) : 281-287 ; "De l'établissement des lignes pour le service international (11ème article). Renseignements complémentaires sur les lignes de l'Italie et de la Norvège," *Journal télégraphique* 20 (1871) : 297-298 ; "De l'établissement des lignes pour le service international (12ème article). Pays-Bas," *Journal télégraphique* 21 (1871): 313-316 ; "De l'établissement des lignes pour le service international (13ème article). Suède et Suisse," *Journal télégraphique* 22 (1871): 329-334 ; "De l'établissement des lignes pour le service international (14ème et dernier article) . Turquie, Wurtemberg," *Journal télégraphique* 23 (1871): 345-348.

83 «L'appareil *Morse* reste provisoirement adopté pour le service des fils internationaux», art. 3, *Convention télégraphique internationale de Paris 1865* ; «Les administrations télégraphiques emploient, autant que possible, des appareils *Morse* marquant des traits colorés», *Règlement de service international destiné à compléter les dispositions de la convention télégraphique de Paris.*

84 «Les Hautes Parties contractantes s'engagent à affecter au service télégraphique international des fils spéciaux, en nombre suffisant pour assurer une rapide transmission des dépêches. Ces fils seront établis dans les meilleures conditions que la pratique du service aura fait connaître. Les villes entre lesquelles l'échange des correspondances est continu ou très-actif seront, successivement et autant que possible, reliées par des fils directs, de diamètre supérieur, et dont le service

drawn up by the French in their capacity as the conference hosts saw the obligation and not the option to adopt thicker wires[85]. However, with the backing of the Spanish delegates, Curchod squashed the two incriminated paragraphs[86] and then introduced a more accommodating disposition, which earned unanimous approval[87].

The question came up again at Vienna, when D'Amico suggested fixing the diameter of international wires[88], which after a brief discussion was set at a minimum of 5 millimetres[89]. It was also decided to consider

---

demeurera dégagé du travail des bureaux intermédiaires», art. 3, *Convention télégraphique internationale de Paris 1865.*

85 «Les conducteurs principaux, formés de fils de gros diamètre, relieront entre elles les capitales des États et les villes d'une importance exceptionnelle. Ils auront pour affectation unique et invariable la communication directe des bureaux placés à leurs extrémités. Des dispositions analogues seront prises, autant que possible, pour les fils secondaires destinés à relier entre elles les villes de moyenne importance. Comme celui des conducteurs principaux, le service de ces fils restera dégagé du travail des petits bureaux situés sur leur parcours», art. 1, *Projet* de la *Convention télégraphique internationale de Paris 1865.*

86 «Sur les deux paragraphes suivants, MM. De Hakar et Curchod font observer que la disposition leur en semble trop impérative. Ils reconnaissent que ces paragraphes posent un principe utile, qu'il est désirable de voir introduire dans le service international ; mais ils ajoutent que ce principe ne saurait être appliqué sans restriction par certains États. M. Curchod propose la suppression de ces paragraphes dans le texte du traité, et leur insertion dans un protocole spécial, comme étant l'expression d'un vœu général et ne constituant aucune obligation», *Documents Diplomatiques de la Conférence Télégraphique Internationale de Paris*, 115.

87 «M. Curchod demande la parole pour revenir sur les deux derniers paragraphes de l'article premier, qui ont été rayés dans le Projet, pour être insérés, comme vœu, dans le procès-verbal des séances. Plusieurs membres ont paru craindre qu'une mesure dont tous les États reconnaissent l'utilité fût cependant négligée, si elle n'était indiquée que dans un document de cette nature. M. Curchod a cherché une rédaction nouvelle pour les deux paragraphes de l'article premier. Il pense que la rédaction suivante pourra réunir tous les suffrages : "Les villes entre lesquelles l'échange des correspondances est continu ou très actif seront successivement, et autant que possible, reliées par des fils directs de diamètre supérieur et dont le service demeura dégagé du travail des bureaux situés sur le même parcours"», ibid., 122.

88 «M. d'Amico propose de fixer le diamètre des fils internationaux», *Documents Diplomatiques de la Conférence Télégraphique Internationale de Vienne*, 209.

89 «M. Jagerschmidt fait remarquer que sur ce point l'uniformité, qui n'est du reste pas nécessaire, ne serait pas toujours possible, le diamètre des fils devant dépendre à la fois de l'étendue et des conditions climatériques de leur parcours. Il considère préférable de déterminer seulement un diamètre minimum. Les diamètres de 4,5,6

suitable for international transmission the Hughes telegraph, which was, for the times, an important technological innovation, the first able to print in the letters of the alphabet. This mechanism had the enormous advantage of making it possible to print messages immediately without using conventional codes such as Morse[90].

The debate had been born when some delegates presented an amendment calling for the Hughes telegraph to be added to art. 3 of the Convention[91]. The Belgian delegate was very much against it:

> M. Vinchent does not think that the Conference has the qualifications to gauge the merit of apparatus: its role must be limited to mentioning those generally consecrated by usage in international service[92].

He felt the duty of the conference was to interpret society's needs and decide to confirm the telegraphs already being used. In other words, the Union was not to impose a standard but value what emerged naturally from society. Not all were in agreement, and in fact some were against using the Hughes because it was too expensive[93] while others preferred to adopt multiple telegraphs of a different type[94]. Substantially, however, many delegates were of the opinion that the conference had the power and the duty to go for the best standard[95].

---

millimètres sont proposés par NN. Falcoïano, do Rego, Comte de Durckheim, de Chauvin et d'Amico. La conférence adopte le diamètre de cinq millimètres comme minimum», ibid.

90 Bontemps, *Les systèmes télégraphiques*, 115-118 ; Bright, *The Electric Telegraph*, 159-160.

91 «l'Article 3 a donné lieu de la part de l'Espagne, de la Hollande, de la Russie et de la Turquie à des amendements dont il est donné lecture par M. le Secrétaire-Général et qui ont tous pour objet l'introduction de l'appareil Hughes dans cet article», *Documents Diplomatiques de la Conférence Télégraphique Internationale de Vienne*, 211.

92 «M. Vinchent ne pense pas que la Conférence ait qualité pour apprécier la valeur des appareils ; son rôle doit se borner à mentionner ceux que l'usage a généralement consacrés dans le service international», ibid.

93 «M. Faber signale le Prix élevé de l'appareil Hughes et du droit de s'en servir, et il considère comme impossible de rendre cet appareil obligatoire. MM. Brändström et Falcoïano se prononcent dans le même sens», ibid.

94 «M. Rubi voudrait y faire figurer aussi l'appareil Schneider qui fonctionne à Vienne de la manière la plus satisfaisante», ibid.

95 «M. Jagerschmidt fait remarquer que le mot "restent" est la constatation d'un fait et correspond à l'état actuel des choses. Il tient d'ailleurs, pour écarter les préoccupations qui ont été manifestées, à bien établir que l'usage de l'appareil Hughes n'aurait rien obligatoire et que les offices conserveraient la faculté de

Three years later in Rome, the Portuguese delegate, backed by Germany and France, suggested modifying article 3 of the Convention and replacing the word "apparatus" with "alphabet". In what appeared to be a formality, the change foreseen was substantial in the approach to technological standardization[96]. With a compulsory alphabet and not a specific kind of apparatus, administrations would be able to choose their telegraphs. Brunner was against the idea and pointed out that there would be a considerable hike in costs for the administrations, which would have to equip themselves with all the apparatus adopted by other states[97], while the chairman added that training personnel to use the new machines would also raise costs[98]. Once again the fear of sending up expenses for technology meant that the delegates hung on to a pre-TU norm.

---

maintenir l'appareil Morse en service. Si la Conférence désirait insister davantage sur ce point, on ajouterait à l'amendement du Gouvernement Impérial et Royal, auquel il se rallie, le mot concurremment après restent adoptés. […] L'amendement du Gouvernement Impérial et Royal est admis par la Conférence, avec l'adition indiquée par M. Jagerschmidt. Il figurera ainsi qu'il suit à la Convention : "Les appareils Morse et Hughes restent concurremment adoptés pour le service des fils internationaux jusqu'à une nouvelle entente sur l'introduction d'autres appareils"», ibid., p. 212

[96] «Au sujet de l'article 3, le Gouvernement portugais propose de substituer le mot "alphabet" au mot "appareils". M. do Rego constate que les appareils télégraphique peuvent changer continuellement par suite de nouvelles inventions et que déjà sur certaines lignes l'on emploie des appareils d'un autre système que les deux appareils approuvés par la Convention, par exemple, le galvanomètre de Thomson ; mais ces appareils prennent toujours pour base les alphabets Morse ou Hughes. M. Meydam appuie l'amendement du Portugal qui sanctionnerait l'emploi des divers système introduits, de fait, sur quelques grandes lignes, tels que l'appareils rapide de Siemens, qui tout en étant différent de l'appareil Morse, en conserve l'alphabet. M. Ailhaud est du même avis. Des appareils différents étant employés effectivement, il convient de mettre la rédaction de la Convention d'accord avec la situation réelle», *Documents Diplomatiques de la Conférence Télégraphique Internationale de Rome*, 252.

[97] «M. Brunner ne verrait pas sans regret toucher à l'article 3, qu'il considère come un article fondamental. Actuellement, chaque État n'est en présence que de deux appareils sanctionnés par la Convention. L'amendement portugais donnerait accès à une multitude indéfinie de système que les relations réciproques des États obligeraient chacun d'eux à introduire, suivant les changements apportés dans tous les pays voisins», ibid., 253.

[98] «M. le Président appuie cette manière de voir et ajoute qu'a cet inconvénient s'en joindrait un autre, celui de la difficulté de former un personnel en état de correspondre avec les divers appareils qui seraient en usage au-delà de chacune des frontières. L'article 64 de la Convention donnant, d'ailleurs, la faculté aux

The same immobility occurred at St Petersburg, where the debate over the apparatus to use on international lines was set off by Curchod reading a letter from Stearns, the inventor of a new system of telegraphic communication[99] based on a phenomenon of physics known to telegraph operators since the origins of telegraphy. Indeed everyone knew "perfectly well that when two offices contended the use of a circuit, the signals of each one were affected by the other's"[100]. Stearns merely understood, as had other scientists, that it was sufficient to find the means to "separate the effect on the needle of the power being transmitted from that of the power being received"[101], in order to decipher messages being sent simultaneously in opposite directions. He conceived for this purpose a special disposition of the circuit between the two sending and receiving mechanisms, and introduced an instrument known as a condenser. Thanks to this system it was possible to separate incoming and outgoing electric signals, thus obtaining simultaneous transmission of the telegrams[102]. The cost of this system, moreover, was not high, since it could be applied to the Morse and Hughes telegraphs.

Faced by Curchod's implicit proposal, the German delegate Stephan was unable to refrain from proposing the alternative multiple system to Stearns' invented by his fellow German, Meyer[103]. The latter had created a new mechanism which, according to the model, was able to send four or six messages simultaneously, but always in the same direction. As stated

---

différents États de prendre entre eux des arrangements particuliers pour l'emploi d'appareils spéciaux sur certaines lignes et dans cas déterminés, il ne voit aucune nécessité de changer l'article 3», ibid.

[99] «Sur l'invitation de M. le Président, M. Curchod donne lecture d'une lettre que le Bureau international a reçue de M. Stearns pour demander que la Conférence mette son Système, qui est appliqué, entre autres, aux États-Unis, dans la Grande Bretagne, en France et en Italie, au rang de ceux que le Règlement reconnait pour la correspondance internationale», *Documents Diplomatiques de la Conférence Télégraphique Internationale de St. Petersburg*, 398.

[100] Robert S. Culley, *Manuale di telegrafia pratica* (Roma: Tipografia Bencini, 1873), Appendix to the Italian edition, 511.

[101] Ibid.

[102] George Bartelett Prescott, *Electricity and Electric Telegraph* (New York: Appleton & C., 1881), 792.

[103] «M. Stephan ne méconnait pas le mérite du système Stearns ; mais il y aurait injustice, à son avis, à ne rendre hommage qu'a un seule inventeur et il prie, en conséquence, les délégués des pays où l'appareil Meyer à transmission multiple a été, croit-il, introduit avec succès, de faire connaître également à la Conférence les résultats qu'ils en ont obtenus», *Documents Diplomatiques de la Conférence Télégraphique Internationale de St. Petersburg*, 398.

by the inventor himself, "the purpose of the system was to allow several employees to exchange a number of telegrams simultaneously on the same wire, [with this mechanism] based on the distribution of time and work"[104]. Meyer's multiple telegraph consisted in four or six receivers paired with separate keyboards, placed on a counter so that a group of operators could simultaneously sit and work at them. Since transmission was made in the usual Morse code, for the operator receiving on paper was just a run-of-the-mill action. What made the Meyer such an unusual telegraph system, however, was a device called the distributor. It consisted in a disc divided into four or six equal parts to which the eight wires departing from each key of each keyboard were connected. Turning on this disc was a device known a carriage consisting in an arm ending in a point called a brush, which followed a synchronized mechanism and made it possible at every turn to connect each of the transmitting systems to a receiver. The device was also provided with a mechanism making self-correction possible in the case of the receivers and transmitters not being perfectly synchronized. The Meyer system was therefore very advanced from a technical point of view[105].

In spite of the number of the administrations speaking in favour of Stearn's or Meyer's systems[106], both were excluded from article III of the new regulation, which ended by cloning the old article 3[107].

---

[104] B. Meyer, "Apparato a trasmissione multipla", *Manuale di Telegrafia pratica* ed. Robert Culley, (Roma: Tipografia Bencini, 1873), 559. Available with the Italian edition.

[105] Ibid.

[106] «MM. Les délégués de l'Italie, de la France et de la Grande Bretagne font connaître que, en effet, le système Stearns, appliqué, dans ces différents pays, avec plus ou moins d'extension, y a donné des résultats très favorables ; mais M. d'Amico ajoute que ce système n'étant qu'un procédé pour augmenter le rendement des appareils Morse et Hughes, et non un appareil distinct, il n'y a pas lieu de le mentionner au Règlement. [...] MM. Les délégués de la France, de l'Autriche et de la Hollande rendent compte des résultats très-satisfaisants que l'appareil Meyer a donnés dans ces pays, en France, avec six transmissions entre Paris et Lyon, et en Autriche et en Hollande, avec quatre. MM. Brunner et Staring se louent beaucoup de l'avoir introduit et font ressortir les avantages très-appréciés des employés que l'appareil Meyer présente par sa simplicité, la facilité de sa manipulation, l'augmentation considérable du rendement des fils et la disposition de son alphabet, qui évite la confusion des signaux et distingue mieux les lettres et les chiffres», *Documents Diplomatiques de la Conférence Télégraphique Internationale de St. Petersburg*, 398-399.

[107] «Les appareils Morse et Hughes restent concurremment adoptés pour le service des fils internationaux, jusqu'à une nouvelle entente sur l'introduction d'autres

The fact that the pre-TU norm of using Morse on international lines was confirmed at St Petersburg though there were far more innovative systems in use on some lines, underscores once again the snail-like speed of the TU when it came to updating. Nevertheless, or because of this, the need for a technical standardization was emerging very clearly in the way numerous telegraph managers desired to establish a professional training common to all officials of the Union administrations.

4.2.3 Profiling telegraph officials

The TU, and the Bureau in particular, made constant attempts from 1868 onwards to encourage directly and indirectly the development of a technical and professional culture common to all telegraph officials in the countries involved. To this end was published the *Journal Télégraphique*, which presented a collection of scientific, technical and administrative articles on practical cases and suggested solutions to practical problems concerning the telegraph service[108]. These articles greatly encouraged comparisons between the techniques and norms adopted in the different countries and led indirectly to attempts to share a common knowledge. Secondly, as we have seen, the Bureau incentivized and added fuel to enquiries on topics concerning the telegraph service.

From the early seventies D'Amico, probably moved by the need to catch up with the more advanced administrations, suggested on at least two occasions some small improvements which would have helped create a common cultural basis for all telegraph managers. In 1870, for instance, he suggested setting up an international competition for the best scientific work on the service[109]. Not only would it have been an ideal moment for verifying the cultural level of the personnel in the different administrations, but also for identifying possible handbooks for training future officials. Some administrations like the Hungarian one were

---

appareils», art. III, *Règlement de service international annexé a la Convention télégraphique de St. Petersburg.*

[108] Some examples of articles published: "Unité de résistance des circuits électriques," *Journal Télégraphique* 1 (1869) : 5-7 ; "Des timbres-télégraphe. Considérations générales," *Journal Télégraphique* 4 (1870) : 33-39 ; "Compte-rendu des expériences faites par les ordres de l'Administration de l'Allemagne du Nord sur l'emploi de quelques piles galvaniques," *Journal Télégraphique* 5 (1870) : 61-64 ; "Nouvelle disposition de la lame vibrante de l'appareil Hughes," *Journal Télégraphique* 8 (1870): 112-114.

[109] ITU-Corr., feuille n. 82/19, 4 juin 1870.

reluctant to back the idea[110], while others like Austria and Russia[111] felt the competition could be the completion of the research and development that all administrations had been carrying out for years.

Because of the highly technical nature of the telegraph service: "each administration is obliged to deal all the time with studying issues concerning the introduction of the most efficacious way of improving telegraph practices"[112]. Yet at the end the usual diffidence prevailed and the competition was left as only a project, which was to have been presented in Rome 1871, but was annulled because of D'Amico's illness.

Nevertheless, the episode which revealed most clearly the growing need to create a cultural basis common to all telegraph engineers and officials was the attempt to establish an international telegraph school. On 14 March 1871, D'Amico sent Lendi, then head of the Bureau, a letter in which he asked to be allowed to present a project for the creation of an international telegraph school at the approaching Rome Conference. D'Amico gave in great detail the reasons behind his proposal:

> Some Telegraph Administrations take their top men from the ranks of their bureaucrats while others prefer to name out of hand civil engineers of the highest level.
>
> The first system has the disadvantage of generally producing mediocre officials, after a long permanence in a tiring, mechanical and routine kind of service, for a bureaucrat does not push the development of intelligence, does not give the habit of command and does not encourage study.
>
> And the second system can have the inconvenience of the top managers lacking the special knowledge of the telegraph service that is not acquired in public high school or university instruction
>
> And though this lack can be filled by technical schools, that I know, there is nowhere an institution that aims exclusively at training electrical engineers for the telegraph service.
>
> Apart from that, it happens that teaching methods are not the same everywhere and there is a lack of uniformity in the procedures adopted by the different Administrations in electro-telegraph experiences.
>
> That said, I think that if beside the international Bureau there were organised a permanent institution for training the top men of the telegraph service, we would have the advantage of satisfying a very real need[113]

[110] ITU-Corr., feuille n. 123/1, 10 janvier 1871.
[111] ITU-Corr., feuille n. 82/21, 1870 (s.d.).
[112] ITU-Corr., feuille n. 123/2, 23 janvier 1871.
[113] «Quelques Administrations télégraphiques prennent les fonctionnaires supérieurs dans les rangs du personnel des bureaux, tandis que d'autres préfèrent nommer, d'emblée, des ingénieurs civils aux grades plus élevés. Le premier système a le désavantage de produire en générale des fonctionnaires médiocres,

Lendi answered only when D'Amico sent another letter as a reminder[114], to explain he was very much against the international school, and listed four reasons:

1. Difficulty in getting a uniform teaching body in a uniform language.
2. Impossibility to provide a complete education for each country and obligation for each country to complete thus teaching with a special second institution.
3. Increased expenses, both for a top class establishment and the annual running costs, at a level which is quite out of proportion to the advantages obtained.
4. Difficulty in choosing a satisfactory seat for the institution for the necessary conditions[115].

In short, Lendi judged as limits the very difficulties that D'Amico had pinpointed as stimuli for opening an international school.

Given Lendi's hostility, his absence from Rome and the sporadic presence of D'Amico, the subject of the school was only taken up in

---

puisque une longue permanence dans un genre de service fatigant, mécanique et routinier, tel que celui des bureaux, n'est pas favorable au développement de l'intelligence ne donne pas l'habitude du commandement et n'encourage pas l'étude.

Et le second système peut avoir l'inconvénient de laisser désirer dans les fonctionnaires supérieurs ces connaissances spéciales du service télégraphique qu'on n'acquiert pas dans les cours publics d'instruction Lycéale [sic] ou Universitaire.

Et quoi qu'on ait pu combler cette lacune moyennant des instituts techniques, on n'a nulle, part, que je sache, une institution qui se propose exclusivement de former des ingénieurs électriciens pour le service télégraphique.

A part de cela, il est arrivé que les méthodes d'enseignement ne sont pas les mêmes partout d'où un manque d'uniformité entre le procédés adoptés par les différents Administrations en matière d'expériences électro-télégraphiques.

Cela posé, je crois que si, à côté du Bureau international, fût organisée une institution permanente pour former des chefs de service télégraphique on aurait l'avantage de satisfaire á un besoins bien réel», ITU-Corr., feuille n. 132/1, 14 mars 1871.

[114] ITU-Corr., feuille n. 132/2, 30 mai 1871.

[115] «1. Difficulté d'un professorat utile en langue uniforme; 2. Impossibilité de donner, en commun, un enseignement complet et obligation pour chaque pays de compléter cet enseignement commun par une seconde institution spéciale; 3. Élévation des frais, tout au point de vue de premier établissement que de l'entretien annuel à un chiffre hors de proportion avec les avantages obtenus; 4. Difficulté de choisir pour l'institution un siège satisfaisant aux conditions voulus» ITU-Corr., feuille n. 133/3, 6 juin 1871.

January 1873, when the administrations' opinions began to arrive at the Bureau[116]. In the April of the same year D'Amico went back into the attack and asked Curchod, by then head again of the Bureau, for an updating on the enquiry[117]. A few weeks later Curchod answered that the enquiry had given a negative result. In fact, only Spain and Russia had spoken for the school while as many as eleven had been against it. Curchod, however, was in favour of D'Amico's proposal and felt that opposition to the school showed the lack of cultural homogenization the school proposed to fill[118]. Differently from Lendi, Curchod recognized at once the educational and cultural differences of the various administrations as a stimulus and not a limit. He suggested reformulating the proposal and transforming the school into an international course to be held, say, at Zurich's Polytechnic, as D'Amico had proposed in the past[119].

After checking that there were no university courses in telegraph engineering on offer in any European country[120] Curchod set to work, always with the enthusiastic backing of D'Amico[121]. He did not write back to D'Amico until October, when he had contacted the Polytechnic in Zurich and had already obtained the potential authorization of the federal government to open there a new course in telegraphic engineering[122]. After

---

[116] ITU-Corr., feuille n. 33/1, 13 janvier 1873 ; ITU-Corr., feuille n. 33/2, 3 janvier 1873 ; ITU-Corr., feuille n. 33/3, 5 mars 1873.

[117] ITU-Corr., feuille n. 33/4, 13 avril 1873.

[118] «Permettez-moi, Mons. Le Dir. Gl., d'ajouter à ce qui précède quelques observations qui me sont du reste, toutes personnelles. Il paraît ressortir des pièces que y'a sous les yeux qu'un assez grand nombre d'Administrations télégraphiques souffrent de l'absence de moyens d'instruction suffisants pour former leur employés supérieurs», ITU-Corr., feuille n. 33/8, 12 mai 1873.

[119] «Aussi ces réflexions m'ont amené à chercher un autre moyen de résoudre la difficulté. Ce moyen consisterait à créer dans un établissement d'instruction supérieur déjà existant, déjà muni de toutes les ressources nécessaires pour les études d'ingénieurs, une branche spéciale pour la télégraphie, comme il y en a pour les ponts et chaussés, pour les mines, pour la mécanique, etc. Si l'on rencontrait un établissement de ce genre qui satisfait unes différentes conditions indispensables (et ici je dois avancer que j'ai songé à l'École polytechnique de Zurich, citée par vous-même dans une de vos lettres sur le sujet), peut être arriverait-on à s'entendre avec les Autorités donc il relève pour y introduire une spécialité nouvelle qui pourrait lui attirer un nouveaux renom et augmenter son importance», ibid.

[120] ITU-Corr., feuille n. 33/5, 9 mai 1873; ITU-Corr., feuille n. 33/6, 9 mai 1873; ITU-Corr., feuille n. 33/7, 9 mai 1873; ITU-Corr., feuille n. 33/9, 18 mai 1873; ITU-Corr., feuille n. 33/11, 1873.

[121] ITU-Corr., feuille n. 33/10, 4 juin 1873.

[122] «J'ai, il est vrai, à la réception de votre communication [...] entretenu de vos idées plusieurs des membres du Conseil Fédéral. Il me semblait en effet nécessaire,

a brief exchange of letters[123], Curchod opened a second official enquiry into the telegraph school, this time asking the administrations for their opinion on the opening of a university course, possibly held in a polytechnic in a neutral territory, like Switzerland[124]. Thanks to Curchod's more detailed project, many states which had been contrary now signalled interest, like Germany, Belgium, France and obviously Switzerland[125]. In spite of some considerable opposition, Norway suggested discussing the project at the next conference, i.e. St Petersburg[126].

There, with the backing of several administrations, D'Amico picked up the main points of his project, which by then had been modified by the ideas of Curchod:

> The Italian Administration thinks that all Offices need telegraph engineers capable of dealing with high level technical questions calling for indepth knowledge. Each administration, it is true, can turn to engineers from other branches of the applied sciences or the bureau heads. But the former are usually without the specific knowledge and the latter general knowledge, both necessary for an electrical engineer. As each state only needs a relatively small number of these figures it cannot organize by itself an establishment for their training. But what it is difficult to do alone, an association of states can bring about to the advantage of all its members [...] Some administrations have doubtless been able to have eminent engineers they have formed themselves, but besides being an exception, it is a result which can only be obtained by means of a hard and difficult effort on their part. It has been said that candidates can be sent to eminent electrical engineers, but an isolated training without organized control is without the necessary guarantees and it could be feared that study time of this nature is not well spent [...] The idea of Italy was to annex to the International Bureau an establishment. But it has found that setting it up at Bern presents many difficulties and it dreams of locating it at the Zurich Polytechnic[127].

---

avant de faire des ouvertures aux autres Administrations de pouvoir compter avec quelque probabilité sur les bonnes dispositions des autorités fédérales à l'égard de notre projet. Les réponses n'ont pas été défavorables [...]. Puis on m'engagerait à parler au Directeur du Polytechnicum [sic.]» ITU-Corr., feuille n. 33/13, 24 octobre 1873.

123 ITU-Corr., feuille n. 33/14, 18 novembre 1873.

124 ITU-Corr., feuille n. 75/1, 13 décembre 1873.

125 ITU-Corr., feuille n. 75/4, 1873

126 ITU-Corr., feuille n. 75/7, 8 mars.

127 «L'Administrations italienne a pensé que tous les Offices avaient besoin de quelques ingénieurs télégraphiques en état de traiter les questions techniques d'ordre élevé demandant des connaissances approfondies. Chaque Administration

Curchod and Hammer, director of the Swiss telegraphs, attempted to defend D'Amico's proposal and turn it to their advantage. In fact, when faced by objections over costs, the languages to use for instruction and the seat of the school, they suggested opening a course at Zurich's Polytechnic. The costs would be low because it would take place at an already existing university. Secondly, language would not be a problem because Switzerland was traditionally multilingual and all would be content with the seat because it would be in a neutral nation which hosted the permanent organ of the Union[128]. At the end of the debate, the project underwent a particular treatment. It was not put to the votes as an amendment to the convention or regulations, but delegates were asked for

---

peut, il est vrai, faire appel aux ingénieurs des autres branches d'application scientifique ou aux chefs intelligents de ses bureaux. Mais les premiers manquent généralement des connaissances spéciales et les autres des connaissances générales, nécessaires, les unes et les autres, à un ingénieur électricien. Comme chaque État n'a besoin que d'un nombre relativement petit de ces fonctionnaires, il ne peut organiser à lui seul un établissement pour leur éducation. Mais ce qui lui est difficile de faire isolément, l'association des différents États peut le réaliser au bénéfice de chaque associé. [...] Quelques Administrations ont pu, sans doute, avoir des ingénieurs éminents qui se sont formés eux-mêmes, mais, outre que c'est l'exception, ce résultat n'a pu être atteint que par un travail beaucoup plus persévérant et plus difficile de leur part. L'on a également dit que l'on pouvait envoyer les candidats auprès d'un ingénieur électricien distingué ; mais cette éducation isolée, sans surveillance organisée, manque des garanties nécessaires et l'on peut craindre que le temps d'études ne soit pas toujours bien employé. [...] L'idée de l'Italie était d'annexer au Bureau international l'institution qu'il s'agissait d'établir. Mais on a trouvé que sa création à Berne présentait bien des difficultés et l'on a songé alors à la placer à l'école polytechnique de Zurich», *Documents de la Conférence Télégraphique Internationale de St Petersburg*, 548-549.

[128] Hammer, for example, stated: «Le Conseil fédéral a soumis la question au préavis d'une commission, composée de professeurs distingués de l'école polytechnique de Zurich. Cette commission a reconnu deux choses: d'abord qu'il n'existait pour le moment aucune institution du genre de celle que l'on voulait établir et, en second lieu, qu'il serait possible de l'organiser comme une section de l'école polytechnique sans frais très-considérables et sans porter atteinte à l'organisation générale de l'école. Cette commission croit nécessaire d'établir un cours de trois années, dont la première serait affectée à l'enseignement supérieur des sciences scientifiques préparatoires et les deux autres à l'enseignement supérieur des sciences physiques et mathématiques, complété par une instruction approfondie de la télégraphie technique et administrative», ibid., 550.

a potential adhesion to the international school[129]. It can be surmised that the chair, Curchod and D'Amico avoided trying for an amendment for fear of the proposal being quashed. Nevertheless, the scarce adhesion to the initiative (only Spain, Italy, Norway, Russia and Switzerland were in favour) put an end to the attempt[130], which was made official some months later in the brief correspondence between Curchod and D'Amico[131]. Yet there was some reaction to Curchod's move, for the Head of the Zurich Polytechnic gave Arnold Tobler the task of holding the first course in telegraphs[132]. Whatever, the projects, inquiries and debates which developed around the subject showed very clearly that many telegraph officials felt very acutely the need to standardize their own culture and practical know-how.

4.5 Conclusions

In a long term perspective the TU's capacity to impose standards needs to be revisited. Its general principles, shaped by liberalism and the norms for communication, were in fact taken on board by conventions signed before 1865. However, in spite of the fact that some delegates signalled the need for changes during the debates, the rules were left unchanged as time passed, thus confirming a conservative inertia as one of the TU's constant features. In a similar way any technological standardizing took the form of recognizing formally technologies which had already been introduced. Initially standards were only fixed on internationally agreed technologies like the Morse code and iron wire, and decisions were avoided over imposing models and materials for batteries, poles or

---

[129] «M. le Président consulte la Conférence pour savoir quels sont les Offices qui seraient disposés à s'associer à la proposition de l'Italie, leur adhésion n'impliquant point, d'ailleurs, un engagement de contribuer, en tout état de choses, à la création de l'œuvre proposée, mais seulement de chercher de concert les moyens de la réaliser», ibid., 551.

[130] «Les délégués de six États. A savoir ceux de l'Espagne, de l'Italie, de la Norvège, de la Russie, de la Suède et de la Suisse, se prononcent pour cette adhésion, les autres États s'abstenant dans la question», ibid.

[131] ITU-Corr., feuille n. 95/1, 7 mars 1876 ; ITU-Corr., feuille n. 95/1b, 6 décembre 1876.

[132] Arnold Tobler initiated the course "Das Telegraphen-u. Signalwesen der Eisenbahnen" in the 7th section. In winter 1878 he continued the experience with the course, "Ausgewählte Kapitel aus dem Gebiete der electrischen telegraphie". See the collection of the Zurich Polytechnic programmes: *Eidgenössiche Technische Hochschule 1855-1955. École Polytechnique Fédérale*, in ETH-Bibliothek Zurich.

insulators. The TU was traditionalist and incapable of answering rapidly and efficaciously to innovations that were coming onto the market, like urgent telegraphs. The expectations were therefore betrayed of those delegates who at Paris had deliberately avoided giving a calendar to the conferences in order to convene meetings when needed to assess innovations and changing circumstances. What most probably happened was that the gradual upsizing of the TU with new members brought in administrations working with more obsolete networks (Russia, Turkey and the Austrian Empire), which therefore slowed down any adoption of new technologies. Attempts to create a shared technological culture via publications, prizes and especially a telegraphic school all went towards filling the gap between the most advanced and developing administrations. But in this case too, conservatism and inertia prevailed over all internal tensions and strivings.

CHAPTER FIVE

A WORLDWIDE PRICING SYSTEM

Though issues over general norms took up most space in the conventions and rules, more words were spent in the 1865-1875 conference debates over tariff uniformity than either standardizing general norms or technology[1]. The considerations determining the calculation of international charges were of utmost importance for two reasons: 1) the revenue from these public administrations went to the state; 2) the governments were responsible to their citizens, and especially the business class[2], for guaranteeing access to a public service at acceptable prices. It followed that a certain ambiguity was caused by the fact that membership of the TU was only allowed to states running the telegraph service. Their administrations were therefore in the position of both running the service and checking on how it was run. A further complication was that while all states enjoyed an income from the international service, the needs of small countries in centrally strategic positions like Belgium and Switzerland were totally different from those of the great empires such as Russia and Turkey, enormous and straddling the margins of TU territory. The differences were such that they brought about an ongoing dialogue, worked out compromises between the interests of the parties involved, often giving rise to thought-provoking discussions, further enriched and complicated by the arrival of Great Britain and the private companies.

Indirectly linked but fundamental for understanding international tariffs was the liberal-influenced decision to route telegrams as cheaply as possible, and the relative debate over an anti-dumping norm which was anything but liberal in inspiration. The three tendencies of international

---

[1] See for example the 1865 Paris: *Documents Diplomatiques de la Conférence Télégraphique Internationale de Paris*, 143-210.

[2] In most of the European states in the period, the governments were nominated by parliaments elected by limited male suffrage excluding the lower classes. In almost all of Europe, therefore, the governments were the expression of the middle and upper classes and government decisions were strongly influenced by their economic interests.

tariff regulations seemed indeed to be fully in line with liberal economic thought: 1) a gradual reduction of charges; 2) the establishment of a pan-European uniform tariff; 3) the gradual introduction of charging per word. They were all proposals which ended by being quashed by the TU's instinctive conservatism and penchant for inaction. It needs to be added, too, that the way in which single national delegations spoke for or against different methods of charging highlights their different interpretations of the telegraph service, which polarised around two contrasting positions of "a service for the few" and "a service for all".

5.1 Routing telegrams and anti-dumping norms

Most telegraph administrations obviously operated like companies, aiming at maximising returns and containing costs. Nevertheless, as within national borders the desire was to safeguard the business class of which the government was often expression, domestic telegrams were cheaper than international ones despite their higher cost. In fact, while domestic telegrams incurred both sending and receiving costs, terminal international ones had only either receiving or sending costs, while if they were transit, the only outlay was transmission[3]. In addition, international telegrams incurred no extra expense for specific infrastructures or specialized personnel, and they used the same operators as the domestic service. It followed that the marginal profit of international telegrams was considerably higher than that of domestic telegrams[4].

---

[3] For example, Anderson, an authoritative voice on the subject, had calculated the cost differential between domestic and international telegrams: «Various methods have been tried to calculate the units of work and of expenses corresponding to each kind of telegram; between the results obtained there have only been insignificant differences, and the mean is expressed in the most simple and practical manner by the following proportions: Interior telegram 5 units of expense, International telegram 3 units of expense, transit 2 units of expense», James Anderson, "Statistics of Telegraphy," *Journal of the Statistical Society of London* 35, no 3 (1872): 272-326.

[4] «Il est notoire qu'un télégramme interne, imposant à l'Administration d'un même pays deux séries d'opérations au moins, au départ et à l'arrivée, occasionne à l'Administration beaucoup plus de frais qu'un télégramme échangé avec l'étranger. Cependant, dans la plupart des pays d'Europe, la parte de taxe afférente à un télégramme international est plus élevée que la taxe interne. Est-ce un mal ? Évidemment non. Le bon marché à l'intérieur rend beaucoup plus de services que n'en rendrait le bon marché avec l'étranger, non seulement parce que le nombre des correspondances est beaucoup plus grand, mais aussi que la distance est moins grande et parce que les circonstances dans lesquelles on correspond avec l'étranger

For a full understanding of the way in which the administrations received income from international telegrams, we need to consider the TU's accounting system. Each administration collected payment for the international telegrams, parts of which were due to all the other administrations involved in its transmission. If a telegram was sent from Italy to Spain and passed through France, the Italian office collected the payment for all the charges of transmission and arrival, which then had to be sent on. The criteria of international accounting therefore served to redistribute the charges to the various administrations involved in the transaction. The method proposed by Vinchent in Paris[5] was employed for many years. It called for debts to be settled only between the administrations of neighbouring states[6], so avoiding accounts having to be settled between administrations with no direct contacts, though at the same time always leaving them the faculty to make other arrangements[7]. Later, in order to make international accounting more accurate, the Vienna Convention authorized the administrations to make bilateral agreements on settling debts with the terminal states along a single telegraph route[8]. In general, however, debts were mainly dealt with by neighbouring states.

---

justifient généralement, de la part de l'expéditeur, une dépense plus élevée sans que cette dépense comprime au même degré le développement du trafic», J. Vinchent, "Considérations générales sur les tarifs télégraphiques (3ème article)," *Journal Télégraphique* 20 (1876) : 368.

5 «Chaque État crédite l'État limitrophe du montant des taxes de tous les télégrammes qu'il lui a transmis, calculées depuis la frontière de ces deux États jusqu'à destination», art. 49, *Convention télégraphique internationale de Paris 1865*.

6 «Chaque État crédite l'État limitrophe du montant des taxes de toutes les dépêches qu'il lui a transmises, calculées depuis la frontière de ces deux États jusqu'à destination. […]Ces taxes peuvent être réglées de commun accord d'après le nombre de dépêches qui ont franchi cette frontière, abstraction faite du nombre de mots et des frais accessoires. Dans ce cas, les parts de l'État limitrophe et de chacun des États suivants, s'il y a lieu, sont déterminées par des moyennes calculées d'après les correspondances d'un mois». […]M. Vinchent propose cet amendement afin d'introduire dans le projet la faculté d'user d'un mode nouveau de comptabilité», *Documents Diplomatiques de la Conférence Télégraphique Internationale de Paris*, 203. What happened was that only the first section of Vinchent's amendment was kept.

7 «Rien ne s'oppose à ce que des groupes d'États adoptent, pour les dépêches échangées entre eux, des arrangements qui simplifient la comptabilité. En convenant, par exemple, que chaque État conservera les taxes perçues, il peut n'y avoir de perte sensible pur aucun d'eux.», ibid., 204.

8 Art. 54, *Convention télégraphique internationale de Vienne 1869*.

The overall profits of the international service were therefore an extremely delicate question for the individual administrations, given their effect on the yearly balance. The greater or lesser revenue involved made the delegates pay close attention to the ways of routing telegrams. Where the international network consisted of a great number of lines, an office could plausibly send telegrams via different routes and if the routes crossed different countries, the charges could well turn out to be different. Initially, in line with the theory of liberalism, the fact that a price differed according to route was not seen as an injustice, as it guaranteed free competition. For example, in 1865 Curchod affirmed that:

> There is no problem in charges varying from one country according to the route taken and that a telegram from Austria to Spain is charged differently if it transits via Switzerland or Italy[9].

On the same subject, the Paris Conference had detailed that in opting for one route or another operators were obliged to go for the cheaper/cheapest one[10]. The French had initially proposed priority for the most direct route, but then they also moved to favour free competition[11]. The general feeling was that as administrations could lower their international charges to attract traffic, it would damage both the clients, who would suffer ill-judged price-cutting at the cost of the service, and the administrations themselves, which would risk a dramatic drop in revenue by running heedlessly after competition. In their dual role of controllers and suppliers, the administrations sought to avoid this double whammy, and later at Vienna drew up an anti-dumping/anti-competition norm, which was tantamount to a cartel agreement.

The norm was put forward the first time under the form of an amendment to article 31 by D'Amico, Chauvin, Lüders, Staring, Vinchent and Jagerschmidt and substantially prohibited price cuts on any already existing route where there was no unanimous agreement between all

---

[9] «il n'y a aucun inconvénient à ce que les taxes d'un pays à un autre variant selon la voie suivie et qu'une dépêche de l'Autriche pour l'Espagne puisse être taxée différemment, selon qu'elle devra passer pour la Suisse ou pour l'Italie», *Documents de la Conférence télégraphique internationale de Paris*, 159.

[10] «La taxe est calculée d'après la voie la moins coûteuse entre le point de départ de la dépêche et son point de destination», art. 37, *Convention télégraphique international de Paris*.

[11] «Le premier paragraphe de l'article 37 est adopté avec la substitution proposée, par M. le Président, des mots "la voie la moins coûteuse" à "la voie la plus directe", qui peut, en effet, ne pas être la moins chère», *Documents de la Conférence télégraphique internationale de Paris*, 189.

involved[12]. The outcome was an impasse between the upholders of free competition like Nielsen[13] and Do Rego[14] and conservatives like Vinchent and Serpos Effendi, the Turkish delegate, who felt an unconsidered cut in tariffs would encourage misuse[15]. The contrast/conflict was embodied in Jagerschmidt, who started out as one of the proposers of the new norm, only to change sides during the debate and speak in favour of a criterion of equality[16]. The evident disagreement between the delegates drew the debate to a halt. Some days later, Chauvin as chairman of a specially appointed committee, gave notice that Curchod's amendment had been chosen from the various ones presented, so once again putting forward a compromise solution that pleased all those who had opposed the preceding formulation[17]:

> The tariffs in the tables can always and at any time be reduced with a common agreement between this or that government involved, but these reductions must have as their aim and effect not to create tariff competition

---

[12] «Toutefois, pour maintenir l'égalité du tarif entre les diverses lignes concurrentes, la taxe applicable à ces lignes ne pourra être réduite sur aucune d'elles, établie ou à établir, sans l'assentiment unanime des Etats traversés par les autres», *Documents de la Conférence Télégraphique Internationale de Vienne*, 313.

[13] «Ces restrictions à la liberté des États sont extrêmement graves; elles modifieraient trop radicalement un des principes essentiels de la Convention, pour que la Conférence, qui est chargée seulement d'un travail de révision, ait le droit de les adopter», ibid., 314.

[14] «M. do Rego cite la concession, faite par le Portugal, d'une ligne destinée à relier ce pays à l'Amérique. La compagnie concessionnaire, qui profitera de l'expérience acquise dans les opérations antérieures de télégraphie sous-marine, n'aura pas besoin, selon toute probabilité, pour assurer la marche régulière de son exploitation, de bénéfices aussi considérables que la compagnie actuelle du câble transatlantique, dont les efforts n'ont été couronnés de succès qu'après plusieurs échecs et des pertes considérables. Forcera-t-on cependant le Portugal à accepter le tarif de cette compagnie ? C'est une mesure qui lui paraît contraire aux intérêts du public et inadmissible», ibid., 313-314.

[15] «Serpos Effendi insiste, de son côté, sur les inconvénients du système actuel qui autorise les concurrences les plus exagérées», ibid., 315.

[16] «C'est cette conséquence que M. Jagerschmidt ne peut accepter, parce qu'elle lui paraît trop contraire aux intérêts des offices concurrents, que l'amendement renferme dans des liens étroits, au moment où leur libre arbitre leur serait le plus nécessaire pour parer aux exigences nouvelles. M. Jagerschmidt veut, dans tous le cas, garantir l'égalité de taxe par les différentes voies», ibid., 313.

[17] «MM. Nielsen et Brändström se déclarent complètement satisfaits de la rédaction nouvelle, du moment où elle laisse aux États entière liberté pour lignes à établir», ibid., 412.

on existing routes, but to open as many routes as possible at uniform charges as equal as possible[18].

The fruit of a compromise, the new regulation, created some problems in interpretation, which were immediately resolved by Vinchent:

> M. Vinchent specifies with the following example the extent of this word [route]. Two cables exist between Belgium and England. If a third one were laid, it would be a line, but not a new route and the prohibition of article 31 would be applicable. Between Norway and Scotland no cable has yet been placed. Telegrams are exchanged via the North of England. If a direct submarine communication is established it will create a new route not subject to prohibition[19].

At the Rome Conference, the anti-dumping norm was quite blatantly turned into a cartel agreement. On opening the debate, Brunner moved in this direction, suggesting leaving the administrations the possibility of coming to agreements on the increase in/ reduction of reciprocal tariffs [20]. At the same time he put forward the idea of abolishing the norm prohibiting tariff reductions to create competition[21]. What Brunner wanted

---

[18]«Les taxes inscrites dans ces tableaux pourront, toujours et à toute époque, être réduites d'un commun accord entre tel ou tel des Gouvernements intéressés, toutefois ces réductions devront avoir pour but et pour effet, non point de créer une concurrence de taxes, entre les voies existantes, mais bien d'ouvrir au public à taxes égales autant de voies que possible» ibid., 411.

[19] «M. Vinchent précise par l'exemple suivant la portée de ce mot [voie]. Deux câbles existent entre la Belgique et l'Angleterre ; si un troisième était posé, il constituerait une ligne, mais non pas une voie nouvelle ; la prohibition de l'article 31 lui serait donc applicable. – Entre la Norvège et l'Écosse aucun câble n'a encore été posé. Les correspondances sont échangées par l'intermédiaire de l'Allemagne du Nord ; si une communication directe sous-marine est établie, elle constituera une voie nouvelle ne tombant pas sous le coup de la prohibition», ibid., 412.

[20] «M. Brunner constate que l'article 34 parle seulement des réductions et ne dit rien des augmentations de tarifs. La question de savoir si la faculté reconnue en matière d'abaissement s'appliquait aux augmentations, a été soulevée, il y a un an, et comme les Administrations n'étaient pas toutes d'accord à ce sujet, cette question a nécessité la réunion de la Commission de Berne. M. Brunner croit qu'il peut se présenter des circonstances où tous les Gouvernements intéressés sont amenés à relever leurs tarifs», *Documents de la Conférence Télégraphique Internationale de Rome*, 385.

[21] «Quant à la disposition qui interdit les concurrences de taxes, elle lui paraît n'être qu'un vœu sentimental, ayant un caractère plus dogmatique que pratique; et,

was to establish a norm allowing cartel agreements also for private cable companies, which were calling for the right to increase and not only reduce their tariffs[22]. An added aspect of the abolition of the anti-dumping norm was that it would favour private companies. They would be put on the same level as public administrations, and end the advantage the latter were enjoying as first movers in the market on many lines. This was why Brunner's second suggestion met the steadfast opposition of D'Amico, who defended the *status quo* in favour of already existing lines and therefore maintaining the public administrations' one step ahead in the market. He, too, was in favour of a cartel agreement, which was why he backed Brunner's first suggestion about the possibility of upping or downing charges, subject to unanimous agreement between the parties. Nevertheless, D'Amico also hoped for public administrations to be one up in respect to the privates. His line, which had already emerged in the way he backed the first Austrian amendment and rejected the second, was accepted by the majority[23], in preference to the conservative position pro keeping the Vienna version of the article[24]. As could be foreseen, the anti-dumping norm was saved. The surprise came over the votes for increasing tariffs – there were as many for as against. In line with the conference

comme dans un traité il faut se borner à inscrire des règles précises et non des principes vagues, il est d'avis de la supprimer», ibid.

[22] «D'autres ne veulent admettre que des réductions. M. Brunner s'étonne que M. le délégué de la Belgique soit au nombre de ces derniers, du moment qu'il veut faciliter les accessions des Compagnies. Pour pouvoir adhérer, celles-ci doivent être laissées libres d'augmenter leurs taxes et on les repousse, en leur refusant une telle faculté. Il considère donc sa première proposition comme étant plutôt une question de rédaction qu'une question de principe», ibid., 388.

[23] «Izzet Effendi adhère aux idées que vient d'exprimer M. le Président. M. Champain appuie la première des propositions austro-hongroises. Les États doivent pouvoir augmenter leurs tarifs, de même qu'ils peuvent les réduire du consentement de tous les États intéressés. Mais il ne croit pas qu'il faille autoriser une concurrence de taxes. [...] M. de Lüders appuie les propositions du Gouvernement austro-hongrois. Entre États européens, il admettrait que l'on ne pût relever les taxes ; mais, pour les communications à grandes distances, l'expérience peut obliger à renoncer à des tarifs trop modérés. Quant au principe de la non-concurrence, il le trouve logique et considère son maintien comme nécessaire, surtout en vues des correspondances avec les Pays lointains», ibid., 386-387.

[24] «Yanco Effendi estime que toute modification doit se faire dans le sens du progrès, c'est-à-dire, de la réduction des taxes. Lorsqu'une Administration a adopté un tarif, elle ne le fait qu'après un mûr examen et il ne saurait, dès lors, être question de le relever. Quant au principe de la non concurrence il croit avec M. le Président, qu'il est très-pratique et doit être maintenu», ibid., 387.

rules, that meant its automatic rejection[25]. But since the debate awoke so much interest, it was re-presented in a later sitting and some weeks later the two amendments were briefly discussed, with the desire to facilitate the emerging submarine cables, particularly in the first case.[26]. This time the first amendment was introduced while the second was again rejected. Thus a norm was ratified which presumed a nascent cartel agreement between public administrations and private companies, though pro the former.

At St Petersburg the anti-competition norm was again attacked. During the works of the tariff committee, the German delegate Stephan held that the regulation had not been applied correctly by national administrations and therefore polemically called for it to be suppressed. He suggested that an alternative could be a slight modification, adding to the prohibition to alter tariffs on existing routes the obligation to keep terminal fares unchanged, whatever the route[27]. The aim here was to stop many administrations from using the particularly common expedient of choosing

---

[25] «La vote donne neuf voix pour l'adoption et neuf contre, M. le délégué de la France s'étant abstenu. En présence de cette parité de voix et conformément à l'article 11 du Règlement des Conférences, l'amendement est considéré comme rejeté», ibid., 391.

[26] «M. le Colonel Robinson appuie de tout son pouvoir le premier des amendements du Gouvernement austro-hongrois. Les différences considérables qui existent entre la position des Gouvernements et des Compagnies lui paraissent nécessiter l'adoption de cette proposition. Un État cherchera toujours, sans doute, à équilibrer ses recettes et ses dépenses, mais un pareil résultat n'offre pas pour lui un intérêt vital comme pour les Compagnies. [...] Il insiste donc pour laisser aux Compagnies autant de liberté que possible, toutes les fois qu'il n'y a pas des rivalités de concurrence. La faculté de modifier les tarifs lui paraît, d'ailleurs, sans danger, du moment que l'accord de toutes les Administrations intéressées est nécessaire pour son application. M. Alan Chambre s'associe aux considérations invoquées par M. Robinson en faveur de l'un des amendements austro-hongrois.», ibid., 579-580.

[27] «Si ce paragraphe vient de suite en discussion, M. Stephan rappellera qu'un amendement de l'Allemagne proposait de le supprimer. Ce n'est pas qu'elle trouvât le principe mauvais, mais c'est parce que dans l'application il est à l'état de lettre morte. Dans la Commission, M. Stephan ayant rencontré chez plusieurs membres le désir de maintenir le principe, a consenti à retirer son amendement, mais en proposant une disposition additionnelle stipulant que les taxes des États terminaux devront toujours rester les mêmes dans les modifications de tarifs faites pour égaliser les voies, addition qui lui paraît indispensable pour que la non concurrence de taxe soit une clause effective. Il ne tient pas d'ailleurs à ce que le nouveau paragraphe amendé soit déplacé», *Documents de la Conférence Télégraphique Internationale de St Petersburg*, 347.

a route with the same total tariff, but offering a greater terminal charge and therefore guaranteeing the sender/receiver office a higher revenue[28].

The new clause was accepted by the majority, though the committee's decision to transfer the whole norm from the Convention to the Regulations set off a debate on its importance. D'Amico was against the move, and stressed that his government had given him the explicit mandate to maintain the norm within the Convention[29]. He managed to win over his colleagues, who voted for the non-competition regulation to be kept within the Convention[30]. His victory was however ephemeral, because two days later the French delegate Ailhaud opposed the modification of the text of the Convention drawn up by the Bureau, which had already been approved by various governments[31]. As modifying the

---

[28] «M. Vinchent croit utile d'ajouter quelques explications complémentaires pour les membres qui n'ont pas assisté aux séances de la Commission. M. le délégué de l'Allemagne a modifié sa proposition primitive, surtout par cette considération que les modifications que vise le principe de non-concurrence sont celles qui se produisent entre deux Conférences. Quand les délégués des États sont réunis en Conférence, ils arrêtent le tableau des tarifs d'un accord unanime et dans toute leur souveraineté, et si cet accord unanime ne s'établit pas, le statu quo est maintenu. Mais ces tableaux une fois arrêtés, il peut y être introduit des modifications de concert entre les Gouvernements intéressés et c'est pour ces modifications que M. le délégué de l'Allemagne admet le principe de non-concurrence, mais avec le correctif de l'égalité des taxes terminales, pour qu'un État, en présence de deux ou plusieurs voies à taxes terminales, pour qu'un État, en présence de deux ou plusieurs voies à taxes égales, n'ait pas un intérêt direct à choisir exclusivement celle qui lui attribuerait une part plus forte», ibid.

[29] «M. D'Amico s'associe aux explications présentées par M. le délégué de la Belgique ; mais il ne saurait admettre que l'on considérât la question de non-concurrence comme secondaire. C'est, à ses yeux, une question essentielle de l'Union des États dont la place ne peut être que dans la Convention», ibid., 348.

[30] «Avant que la Conférence se prononce sur le maintien de la disposition dans le Règlement ou sur son insertion dans la Convention, M. D'Amico doit faire connaître que ses instructions ne l'autorisent pas à consentir à la suppression dans la Convention du principe essentiel de la non-concurrence. La proposition de M. le délégué de l'Italie est adoptée par 13 voix contre 6», ibid., 350.

[31] «M. Ailhaud désirait provoquer une explication que les modifications apportées à l'article 10 du projet de Convention lui paraissent rendre nécessaire, à savoir si la Convention qui sera signée à St-Pétersbourg aura un caractère d'immutabilité ou si elle continuera, comme précédemment, à être soumise aux révisions périodiques des Conférences. Le Gouvernement français était disposé à admettre la première manière de voir, sur la base des projets préparés par le Bureau international ; mais, si l'on introduit dans le texte conventionnel des dispositions qui, à toutes les

non-competition norm changed the text for which the diplomats were mandated, he asked for it to be transferred to the Regulations. The transfer would not have made the norm less important or affected its obligatory nature, but simply made it subject to possible future revisions[32]. This was what Vinchent had felt at the beginning when he had asked D'Amico to reconsider his position[33].

D'Amico decided not to block the flow of the proceedings, and so after explaining his reasoning, accepted the proposed modification[34]. The administrations agreed to maintain the cartel, but also reserved the right to modify or annul it in the future. What it meant was that the cartel had formally lost its importance as it had been transferred from the permanent document of the convention to the more secondary regulations, though it was left intact.

5.2 The evolution of the international tariffs

5.2.1 Uniforming tariffs

At the 1865 Paris Conference the delegates obtained an extraordinary result in introducing the uniform tariff and abolishing different zone payments. The uniform tariff meant charging at the same level any telegram exchanged between two states, whether they housed the sending or arrival office[35]. Before the Paris Conference, all the conventions had provided for charges based on the distance covered by the telegram. The

---

réunions précédentes ont été l'objet de discussions et de changements, M. Ailhaud serait obligé de prendre de nouvelles instructions», ibid., 354

[32] «M. d'Engelhardt répond que dans la pensée du Gouvernement impérial les articles de la Convention doivent avoir un caractère d'immutabilité. Si donc une disposition est de nature à donner lieu ultérieurement à des discussions, il conviendrait de l'insérer de préférence dans le Règlement, dont les prescriptions sont obligatoires au même titre que celles de la Convention et qui n'en diffèrent, comme autorité, que par cela qu'elles peuvent être révisées par les Conférences ultérieures», ibid.

[33] «M. Vinchent demande à M. le délégué de l'Italie s'il ne consentirait pas, en basant sur les déclarations de M. le délégué du Ministère des affaires étrangères à provoquer de nouvelles instructions lui laissant plus latitude à cet égard», ibid.

[34] «M. D'Amico ne se refuse pas, d'ailleurs, à faire les démarches nécessaires pour obtenir, sur ce point, pleine liberté d'adhérer à la décision définitivement prise», ibid., 355.

[35] «La taxe applicable à toutes les correspondances échangées, par la même voie, entre les bureaux de deux quelconques des États contractants sera uniforme», art. 30, *Convention télégraphique internationale de Paris*.

idea of a uniform charge had been inserted as an option in the preliminary project drawn up by the French, but it was only thanks to Curchod's amendment that it became the corner stone of the TU's tariff norms:

> The tariff applicable for all correspondence exchanged on the same route between the offices of any two to the signee states will be the same[36].

The whole text of article 30, as the French admitted, had been taken from article 18 of the Berne Convention, with the precise purpose of finding as wide a consensus as possible[37]. Whatever the case, the new norm was formulated in an extremely complex way and ran the risk of being misinterpreted[38]. Curchod skilfully presented a new article 30 which made very few changes to the original and its compromising nature, but streamlined the form and made it of easy application[39]. He focused on three technical features he found essential: 1) the number of territorial divisions, with two for large empires like Russia and Turkey; 2) the charge to oscillate between a maximum and minimum level; 3) the monetary basis to be shared and possibly represented by the franc[40]. After he had introduced these conditions - the second of which was abolished in the next sitting - Curchod specified that the assembly would have to give further discussion to three other technical details: 1) the monetary basis of

---

[36] «la taxe applicable à toutes les correspondances échangées par la même voie entre les bureaux de deux quelconques des États contractants sera uniforme», ibid., 157.

[37] «M. le Président fait observer que le Projet ne pose là qu'un principe général, devant servir de point de départ aux arrangements particuliers, mais qui n'est point rigoureux au point de s'opposer à des concessions réciproques, lorsqu'on en viendra à ces arrangements», ibid., 152.

[38] «M. Vinchent demande alors quelques explications, notamment sur la phrase finale du deuxième paragraphe, d'après laquelle la parte de chaque État dans la taxe totale doit être fixée en raison du parcours moyen des correspondances sur les lignes respectives de chacun d'eux. Il pense que cette disposition serait trop rigoureuse et qu'elle semblerait s'opposer, de la part des grands États, à des concessions que des considérations diverses pourraient rendre justes et nécessaire», ibid.

[39] «M. Curchod rend hommage aux dispositions libérales du Projet présenté par la France, mais il croit nécessaire d'apporter certaines restrictions à la latitude donnée aux États contractants», ibid.

[40] «Il [M. Curchod] propose un amendement dont les bases sont les suivantes: 1) le nombre des divisions territoriales auxquelles s'appliquerait une taxe distincte ne dépasserait pas deux par État; 2) la taxe varierait entre un minimum de cinquante centimes et un maximum déterminé; 3) la base monétaire serait le franc, et l'on n'admettrait point de fraction inférieur á cinquante centimes», ibid.

an ordinary telegram; 2) the method for establishing charges in the different states; 3) the maximum number of territorial divisions corresponding to the different charges[41]. The following discussion on tariffs concentrated on these three points, showing the level of influence Curchod already wielded.

He proposed taking as the minimum charge unit a single twenty-word telegram, adding for the next ten half of the standard charge and so on[42]. The only opposition came from the Spaniard Hakar, who spoke in favour of an ordinary telegram consisting of ten words[43]. Backed by Vinchent[44], Curchod was against this counter-proposal and held that the first unit of words had to contain double the following units both to incentivise the sender and remunerate the office for the initial cost of transmission. Vinchent felt that the twenty-word telegram had by then been accepted by public opinion and consecrated by a fifteen-year experience[45].

In his amendment Curchod had then specified that all the states would have to decide on their reception and transit charges in agreement with

---

[41] «M. Curchod croit aussi que le Projet doit être plus explicite sur l'établissement des tarifs, et, après avoir montré comment on procéderait en Suisse, il indique le but à atteindre par la Conférence. Elle doit, selon lui: 1° déterminer la base monétaire de la dépêche simple; 2° indiquer la manière dont se fera l'établissement des tarifs entre les divers États; 3° Fixer le nombre maximum des divisions territoriales auxquelles correspondront des taxes distinctes», ibid., 153.

[42] «Le minimum de la taxe s'applique à la dépêche simple, dont la longueur ne dépasse pas vingt mots. Pour chaque dizaine de mots ou fraction de dizaine au-dessus de vingt mots ou fraction de dizaine au-dessus de vingt mots, il sera perçu une taxe égale à celle de la moitié de la dépêche simple», ibid., 158.

[43] «L'Espagne serait disposée à prendre pour unité de dépêche le groupe indivisible de dix mots», ibid., 151.

[44] «M. Vinchent défend l'unité actuelle de dépêche, au même point de vue de simplification. Il estime aux neuf dixième des dépêches celles dont la longueur approche de vingt mots ; ce fait lui semble permettre de négliger entièrement la longueur de la dépêche dans l'échange des comptes internationaux, et c'est pour lui un nouveau motif de conserver le minimum de vingt mots», ibid., 162.

[45] «M. Curchod donne ensuite les raisons qui lui font repousser la dépêche de dix mots comme unité. Il ne comprendrait pas que la première série fût égale aux séries suivantes. Le fait seul d'être admis à correspondre mérite de la part de l'expéditeur une rémunération spéciale, indépendante du nombre des mots et correspondant en quelque sorte aux frais généraux du service. On atteint ce résultat, sans préjudice pour le public, en exigeant pour la première série une double taxe et en accordant, en échange, un nombre double de mots. M. Curchod craindrait d'ailleurs de toucher à une règle qui lui paraît acceptée par l'opinion et consacrée par une expérience de quinze années», ibid., 161.

neighbouring countries[46], not speaking of the way in which single national charges would be put together in a general table. Two possibilities were left open: (1) the states would have to decide on their own charges during the conference so as to attach the table to the convention; 2) the states could decide on their own tariffs after the closure of the conference by means of multilateral and bilateral agreements. Curchod was very much in favour of resorting to multilateral and bilateral agreements, which he felt would give a greater freedom to the single administrations[47]. The main backer of the first solution, the Prussian De Chauvin, suggested organizing three territorial committees there at the conference (northern, southern and western states), each tasked with drawing up the table of charges for its area[48].

The clash between the two visions was resolved yet again via mediation. The charges would be decided individually by each administration and then approved during the conference, not in the committees suggested but in a wider meeting, in order to encourage discussion[49]. In a following sitting the French chairman and the Belgian

---

[46] «Le taux de la taxe sera établi d'État à État, de concert entre les Gouvernements extrêmes et les Gouvernements intermédiaires. Chacun des États contractants, après d'être ainsi entendu sur l'application des bases qui précèdent, communiquera à chacun des autres États contractants la série des taxes uniformes applicables à la correspondance internationale, à partir des frontières d'entrée sur son territoire jusqu'aux bureaux des États ou des Administrations qui auront adopté les principes contenus dans la présente Convention», ibid., 158.

[47] «En ce qui touche le taux de la taxe, M. le Délégué de la Suisse ne croit point possible de le fixer autrement qu'entre les États intéressés. Il insiste pour que le principe de la liberté soit maintenu dans le traité général, tout en désirant voir s'établir une entente immédiate entre les États. Mais la détermination des tarifs exige à ses yeux l'examen contradictoire de questions très-diverses, qui ne pourraient point se poser dans une réunion de tous les États», ibid., 161.

[48] «M. le colonel de Chauvin appelle l'attention de MM. les délégués sur l'extrême importance de la question des taxes et l'opportunité de régler complètement cette question pendant les réunions de la Conférence. Il pense qu'il ne suffit pas de fixer des principes généraux, mais qu'il faut aussi arrêter les tarifs. Il indique la marche à suivre pour résoudre promptement la question. Les États de l'Europe peuvent se classer en trois groupes dont les intérêts sont plus intimement liés : 1) les États du Nord, comprenant la Russie, les États Scandinaves et la Prusse ; 2) les États du Sud, tels que la Turquie, la Grèce, l'Autriche et l'Italie ; 3) enfin, les États de l'Ouest. Ces groupes fixeraient d'abord isolément, et sans tenir compte des autres, l'ensemble de leurs tarifs ; puis ces tarifs seraient discutés en commun. Les tarifs de ces groupes pourraient alors être augmentés ou diminués, mais toutes les taxes du groupe conserveraient les rapports établis par le premier travail», ibid., 173.

[49] Ibid., 176-177.

and Swiss delegates decided that each delegation would draw up a tariff table during the assembly[50]. On 8 April 1865, after several attempts and a twenty-day debate[51], the tables were approved by the conference and added to the convention[52].

Although the Paris Convention had provided for the introduction of a uniform tariff, both the original French proposal and Curchod's amendment allowed for the exceptional admission of two zones in the Russian and Turkish empires[53]. As Curchod himself specified, it was an exception to the prevalent single charge, introduced in order not to put at a disadvantage states covering a larger territory than most of Europe and for this reason severely penalized by having to work with a uniform tariff within their borders[54]. So with the exclusion of Turkey and Russia, each state proposed a transit and destination tax. In general, the latter was higher than the former, but though the Belgian delegates argued hard and fast, they were not able to push through a regulation fixing an objective relationship between the two charges. The Belgian Fassiaux had for instance proposed an amendment setting the transit charge at half that of the destination[55], and the Austrian Brunner followed the same line,

---

[50] «Après un court échange d'observations entre M. Le Président, MM. Faissaux et Curchod, il est convenu que sans interrompre le cours des travaux de la Commission, MM. les délégués s'occuperont, d'abord individuellement, puis de concert, de la préparation du tableau des taxes qui doit, aux termes de la proposition de M. le colonel de Chauvin, figurer comme annexe au traité», ibid., 187.

[51] Ibid., 207; ibid., 223.

[52] Ibid., 294-298.

[53] «La taxe applicable à toutes les correspondances échangées par la même voie entre les bureaux de deux quelconques des États contractants sera uniforme. Un même État pourra toutefois être subdivisé, pour l'application de la taxe uniforme, en deux grandes divisions territoriales au plus. Les États contractants se réservent d'ailleurs toute liberté d'action à l'égard de leurs possessions ou de leurs colonies situées hors des limites de l'Europe», ibid., 158.

[54] «Il [M.Curchod] pense que la Convention doit laisser la faculté de subdiviser les États, car cette disposition est consacrée par l'usage, et les conditions géographiques de certains pays légitiment des exceptions à la taxe uniforme. Dans le cas où l'on pourrait prendre une taxe très petite, un franc par exemple, pour toute l'Europe, il ne serait pas nécessaire d'établir de divisions ; mais la taxe serait à l'avantage des grandes distances et au détriment des petites, qui cependant fournissent la correspondance télégraphique la plus active et la plus importante», ibid, 153.

[55] «M. Fassiaux insiste : il faut, selon lui, décider d'une manière générale quel sera pour chaque État le rapport des taxes de transit avec les taxes réclamées par cet État pour les correspondances en provenance ou à destination de ses bureaux. Il

upholding the need to fix the transit charge lower than the destination one[56].

But France had no intention of reducing the tariffs it had fixed with neighbouring countries, and so accused Fassiaux and Brunner of being illiberal for wanting to impose limits on fixing charges[57]. Fassiaux had actually explained his line of thought, that transit telegrams had no delivery or reception expenses, cost less than destination ones and therefore could be priced lower[58]. Though France was not against his reasoning, it went on opposing any formal regulation of the relationship between the two tariffs, and therefore Austria and Belgium abandoned their amendment with the tacit agreement that all would have to propose lower transit than destination charges[59].

---

estime que ce rapport doit être tout au moins l'égalité. Il cite les vœux exprimés dans le dernier Congrès postal, au sein duquel on a reconnu qu'en matière de poste la taxe de transit ne devrait point dépasser la moitié de la taxe intérieure. Il sait qu'à cet égard le télégraphe n'est pas dans les mêmes conditions que la poste ; mais c'est en tenant compte de ces différences qu'il propose l'égalité», ibid, 164.

[56] «M. Brunner appuie l'opinion de M. Fassiaux. Il fait remarquer que la France est la seule voie entre l'Espagne et le reste de l'Europe, comme l'Autriche, entre l'Occident et la Turquie ; il propose d'ajouter à l'article, après le deuxième paragraphe : «La taxe de transit pour un État quelconque ne sera jamais plus élevée que la taxe uniforme adoptée pour le service international des bureaux de cet État», ibid., 165.

[57] «M. Jagerschmidt s'attache alors à établir qu'il n'existe pas une corrélation nécessaire entre le prix du transit au travers d'un État et celui de la correspondance à destination ou en provenance des bureaux de cet État. Ce dernier prix doit être fixé d'après la moyenne des distances parcourues, tandis qu'entre deux frontières il n'y a qu'une seule et même distance. La taxe de deux francs a été, vis-à-vis des États limitrophes de la France, l'expression de cette moyenne. Appliquée uniformément au transit de ces États entre eux, elle constituerait au profit de quelques-uns des avantage nouveaux, qui ne sauraient répondre qu'à de nouvelles concessions. M. Jagerschmidt fait ensuite remarque que, tant que n'existent pas les communications directes dont l'article 1ère du Projet porte le germe, le transit donne lieu plus souvent à un travail plus considérable que la correspondance entre deux États», ibid., 184.

[58] «Il [M. Fassiaux], constate que cette manière de voir n'est point partagée par la France, mais il croit que c'est là une erreur, et il ne comprend pas qu'une dépêche remise à domicile à la frontière soit, pour un État, moins onéreuse qu'une dépêche transmise par le télégraphe au-delà», ibid.

[59] «M. Brunner demande alors la parole : il avait regretté pendant temps d'avoir provoqué, par sa proposition, la discussion qui occupe encore la Conférence ; il ne le regrette plus aujourd'hui, en présence des déclarations si libérales de la France. Il est maintenant convaincu qu'on doit arriver à une entente facile, et qu'il devient

To sum up, the Paris Convention decided that a twenty-word telegram was to be priced the same as the sum of the destination charges of the sender and destination states plus all the transit charges of the countries it crossed.

Beyond these results, a far more revolutionary move, the introduction of a single charge, was discussed at Paris. It was first advanced by Hakar, who proposed applying the same destination and transit tariffs to all states in the TU and establishing a kind of telegraphic postage/franking[60]. A reform like that was indeed too radical for most of the delegations, but the Spaniards insisted and kept on re-proposing their amendment which, as was to be expected, was rejected[61]. A second project for a single tariff was presented by the French, who had not originally even foreseen a uniform charge as compulsory. The chairman of the sitting took advantage of the general disarray on the formulation of the articles on tariffs and at the end of the sixth sitting presented a note rather than a formal proposal, in which he spoke of the introduction of a single six-franc charge for all telegrams exchanged among the European states[62].

---

inutile d'inscrire dans les traités des principes qui paraissent arrêtés dans l'esprit de tous. Il croit devoir dès lors retirer entièrement sa proposition. M. Fassiaux s'associe à la déclaration de M. le Délégué de l'Autriche», ibid., 185-186.

[60] «M. de Hakar expose les principes généraux sur lesquels MM. les Délégués de l'Espagne baseraient l'établissement des tarifs européens et la taxe uniforme. Il leur paraîtrait convenable d'appliquer dans tout les États la même taxe au même nombre de mots: Les différences existent aujourd'hui dans les taxes conduisent à des complications de comptabilité qu'il serait utile de faire disparaître.», ibid., 151.

[61] «M. de Hakar combat les conclusions de M. Curchod, qui lui paraissent maintenir le système des zones, et, reprenant la pensée développée au début de la discussion sur l'article 30, propose pour conclusions l'amendement suivant : «Dans tous les pays, le même nombre de mots aura la même taxe, de sorte que la taxe devra se multiplier autant de fois qu'il aura de pays à traverser»», ibid., 153 ; «M. de Hakar se déclare de nouveau opposé à ces principes mêmes. Il estime que c'est se placer encore sur le terrain des Conventions de Bruxelles et de Berne. Il irait, lui, jusqu'à consentir des réductions proportionnellement plus fortes que celles des autres États, pour arriver au résultat qu'il désire, c'est-à-dire à l'établissement d'une même pour chaque État», ibid., 184.

[62] «M. le Président présente alors la note suivante, sur laquelle il appelle toute l'attention de la Commission, sans en faire toutefois l'objet d'une proposition formelle: "L'adoption de tarifs uniformes et réduits n'a rencontré dans la Conférence aucun dissentiment. La plus large application du principe généralement admis consisterait à établir une taxe unique pour les dépêches échangées entre les divers États de l'Europe, […] Partant de cette donnée, il semble que le chiffre de six francs pourrait être admis comme taxe internationale européenne. […] La proposition précédente pourrait être libellée comme il suit : la taxe d'une dépêche

The French had no more success than the Spaniards with most of the delegates. Given their difficulty in reaching a common agreement on tariff uniformity, it is scarcely surprising that they were so diffident about a standard charge. And then there was always a kind of primordial fear that a single charge would be a step towards national administrations losing autonomy and finding their sovereignty being limited.

5.2.2 Outside Europe

The method for charging established in Paris was kept unchanged for European correspondence but from 1868 extra-European tariffs were modified. At Vienna the number of words in ordinary telegrams bound out of Europe was cut from twenty to ten while at Rome charging per word was introduced.

At the 1868 Vienna Conference D'Amico[63] and Falcoïano[64], were the first to suggest reducing the number of words to 15 and 10[65]. They would have liked to reduce the unitary cost of the international telegram, but met a compact opposition. Nevertheless, at the end of the brief debate the British India delegate Goldsmid proposed reducing the word count of an ordinary telegram to India, which was dearer than others and was given special treatment[66]. Vinchent's answer was to suggest a ten-word limit to

---

simple échangée entre deux États de l'Europe non contigus est fixée à six francs. Sur cette taxe, une somme de un franc ou de cinquante centimes, suivant l'étendue du pays traversé, sera dévolue pour prix du transit à chacun des offices intermédiaires. L'excédant sera partagé, d'après la même base, entre les offices expéditeur ou de destination"», ibid., 178-179.

63 «Cet article a fait l'objet d'un amendement de l'Italie. Cet amendement, modifié en séance par M. D'Amico, est ainsi conçu: "Fixer la dépêche simple à quinze mots en réduisant aussi le tarif actuel, et augmenter la taxe du tiers pour chaque série indivisible de cinq mots au dessus de quinze"», *Documents de la Conférence Télégraphique Internationale de Vienne*, 293.

64 «Falcoïano propose de fixer le minimum de la dépêche simple à dix mots et la progression de la taxe de dix en dix mots. Cette combinaison, qui répondrait au vœu de l'office des Indes, aurait encore pour résultat de simplifier le calcul du nombre des mots et l'application de la taxe», ibid.

65 «M. D'Amico expose les motifs de cet amendement, qui aurait pour effet: 1° de décharger les lignes d'un travail inutile, les expéditeurs mettant souvent dans leurs dépêches des indications superflues, afin d'épuiser leur droit jusqu'à vingt mots», ibid.

66 «M. le Colonel Goldsmid, revenant sur l'amendement de M. D'Amico, en approuve complètement le principe. Il y aurait grand intérêt pour le public, à raison du prix élevé des correspondances avec les Indes, à lui accorder la dépêche de dix

all extra-European correspondence. He was aiming at extending a possible tariff advantage to maritime cable companies and at the same time limit the use of "packaging messages" by private companies, which managed to cut costs with their private codes compacting several messages into an ordinary 20 word telegram. With a 10 word limit in a telegram Vinchent hoped to reduce the acts of fraud carried out against the telegraph administrations[67]. Most of the delegates approved of the proposal, which was then introduced into the new convention[68].

In the following Rome conference (1871), the debate on tariffs also concentrated on charges for extra-European telegrams. Once again the debate was opened by D'Amico, who proposed cutting the number of words down to 15, quoting the positive results of the Italian experience of a word reduction leading to a radical increase in the number of telegrams sent[69]. D'Amico's proposal came up against the conservatives, who

---

mots avec taxe de moitié. Cette faculté précieuse, surtout pour les correspondances de familles auxquelles peu de mots suffisent, contribuerait à en augmenter le mouvement. Il propose donc "de réserver la faculté, si ultérieurement la nécessité s'en fait sentir, de fixer le minimum d'une dépêche à dix mots pour les correspondances hors d'Europe, en l'assujettissant à une taxe égale à la moitié au moins de la taxe actuelle"», ibid., 294.

[67] «M. Vinchent est d'avis de ne pas borner seulement à la correspondance des Indes l'exception proposée par M. le colonel Goldsmid, mais d'en accorder le bénéfice aux grandes compagnies sous-marins, notamment à la compagnie du câble transatlantique. La dépêche de dix mots est pour cette compagnie une condition nécessaire de son exploitation, non-seulement à cause du taux forcément élevé de son tarif, mais encore pour éviter la concurrence redoutable de certaines agences, qui acceptent de divers expéditeurs des dépêches d'un petit nombre de mots, les groupent de manière à en former des dépêches de vingt mots et les distribuent soit en Amérique, soit en Europe. Le public n'aurait plus intérêt à recourir à ces agences, s'il pouvait composer ses correspondances d'un petit nombre de mots et payer dans la proportion du service qui lui serait rendu», ibid., 295.

[68] «Le minimum de la taxe s'applique à la dépêche dont la longueur ne dépasse pas vingt mots s'accroit de moitié par chaque série indivisible de dix mots au dessus de vingt. *Toutefois les offices à admettre sur leurs lignes la dépêche de dix mots avec taxe réduite, cette dépêche étant d'ailleurs taxée pour le parcours européen comme une dépêches de vingt mots*», art. 32, *Convention télégraphique internationale de Vienne 1868.*

[69] «Quant à l'amendement italien, il aurait pour objet d'abaisser à quinze le nombre des mots de la dépêche simple, et augmenter la taxe de moitié par série de cinq mots supplémentaires. M. le Présidente rappelle que l'Administration italienne a déjà introduit dans son service intérieur la dépêche de quinze mots et, bien que la

appealed to presumed differences in treatment between international and national systems[70], opposed the amendment and caused it to be rejected[71]. Rejected it was, but again it served to call the attention to the extra-European regime. As representative of the *Compagnie du câble transatlantique* Curchod then proposed charging per word once the ten-word limit had been reached, hoping to adapt the convention's regulations to the charging system used by the submarine companies in order to facilitate their joining the TU[72]. The reaction was sharp from the British India delegates, who feared unfair competition from the privates if they could adopt a cost per word system instead of state tariffs[73]. It was

---

mesure soit encore récente, il ne peut qu'en constater les excellents résultats», *Documents de la Conférence Télégraphique Internationale de Rome*, 362-363.

[70] Vinchent explained this concept in great detail in a series of articles published in the *Journal Télégraphique*: J. Vinchent, "Considérations générales sur les tarifs télégraphiques (1ère article)," *Journal Télégraphique* III, no 18 (1876): 327-333 ; J. Vinchent, "Considérations générales sur les tarifs télégraphiques (2ème article)," *Journal Télégraphique*, III, no 19 (1876): 347-351 ; J. Vinchent, "Considérations générales sur les tarifs télégraphiques (3ème article)," *Journal Télégraphique* III, no 20 (1876) : 367-371 ; J. Vinchent, "Considérations générales sur les tarifs télégraphiques (4ème article)," *Journal Télégraphique* III, no 22 (1876): 403-409. Vinchent drew largely from Anderson's already quoted pamphlet: Anderson, *Statistics of Telegraphy.*

[71] «D'un autre côté, M. Vinchent ne considère pas le mot en télégraphie comme une sorte de marchandise que l'on doive payer au prorata de la consommation. Il croit, au contraire, qu'il y a un intérêt considérable à augmenter la proportion des dépêches de vingt mots, pour les employés dont le travail est simplifié du moment qu'il s'opère sur des éléments de même valeur et pour les Administrations qui échangent entre elles des unités de même taxe. Il se prononce donc pour le maintien de la limite de vingt mots», ibid., *Documents de la Conférence Télégraphique Internationale de Rome,* 364.

[72] «Au sujet de l'article 32, M. Curchod désirait soumettre une question à la Conférence. Les Compagnies transatlantiques ont, comme on le sait, adopté pour unité la dépêche de dix mots avec gradation de la taxe par chaque mot supplémentaire. Il va sans dire que pour le parcours européen l'on calcule sur la base de vingt mots, avec progression par dix mots. M. Curchod désirait savoir si ce mode de taxation rentre dans les prescriptions du dernier paragraphe de l'article 32. S'il en était ainsi, il n'y aurait pas sur ce point d'obstacle à l'adhésion des Compagnies. Dans le cas contraire, il serait obligé de demander en leur faveur une dérogation qui leur permît de conserver le système actuel. Il s'agit de taxes très-élevées. La mesure est donc avantageuse pour le public et pour les Administrations à qui l'on ne réclame que le prix rigoureusement proportionnel au service rendu», ibid., 364-365.

[73] «M. Andrews rappelle que MM. Les délégués des Indes ont demandé déjà que la dépêche de 10 mots de fût pas bornée au parcours extra-européen. Les compagnies

however just a misunderstanding, for as Vinchent explained, all convention regulations were applicable to all signees, whether private companies or state administrations[74]. So the British India administration could take advantage of the charge by word system and therefore freely compete with its rival private companies. In spite of this reassurance Vinchent had to furnish another formal explanation for charging per word and he explained that it was an optional measure applicable to both state administrations and private companies[75]. Though the amendment did not

sous-marines unies ont sur cette question formulé une proposition dont l'adoption aggraverait les inconvénients résultant déjà de la situation actuelle. Elles voudraient sur leurs lignes la gradation des taxes par mot ou du moins par cinq mots. Tandisque la progression par dix mots resterait appliquée sue le réseau des États. Une pareille disposition rendrait toute concurrence impossible pour les lignes des Gouvernements», ibid. 365.

[74] «Il [M. Vinchent] croit que la Conférence doit se prononcer sur la question de la gradation par mot ou par dizaine de mots. A Vienne, l'on s'est trouvé en présence d'une concession déjà accordée par deux États à une Compagnie privée et lui reconnaissant la faculté d'accepter des dépêches de dix mots avec taxe réduite. [...] Mais ce qui importe, c'est de conserver le principe de la non-concurrence des taxes. Sous ce rapport, M. Vinchent ne fait pas de distinction entre les États et les Compagnies, et ne se préoccupe que de leur situation géographique. Si les lignes des uns ou des autres sont en Europe, c'est-à-dire si elles sont situées sur le territoire d'États européens ou si elles relient deux points situés en Europe, elles ne peuvent introduire, sur ce parcours terrestre ou sous-marin, la dépêche de dix mots. Si les unes ou les autres sont hors d'Europe, elles ont également, les unes et les autres, le droit de profiter de la faculté que leur confère le dernier paragraphe de l'article 32. Ceci posé, M. Vinchent croit qu'il est d'un grand intérêt pratique de donner au public la faculté d'atténuer la charge de certaines taxes, de même qu'il est très-important pour les grandes entreprises sous-marines, dont les câbles ne donnent aux signaux qu'un écoulement peu rapide, d'utiliser de la manière la plus productive leurs moyens de communication. Les grandes lignes à taxes élevées ont, en outre, à lutter contre le groupement des dépêches. Ces considérations analogues, en adoptant la dépêche de dix mots sur les parcours extra-européens», ibid., 366.

[75] «Il [M. Vinchent] ne comprend pas, d'ailleurs, comment le vote précédent qui autorise simplement l'exercice d'une faculté peut s'interpréter dans le sens d'une obligation. Puisque des incertitudes et des malentendus paraissent s'être produits à cet égard, il précisera la portée de la disposition adoptée par la Conférence. Il demeure, d'abord. Bien établi que sur les parcours européens le régime de la dépêche de vingt mots et de la progression par dix mots demeure intact; par conséquent, toutes les lignes entre l'Angleterre et les voie de la mer, ont, depuis le point de départ jusqu'au dernier bureau européen qu'elles touchent en passant, un parcours européen sur lequel les règles générales sont exclusivement applicables. En second lieu, la faculté de la gradation par mot, sur les parcours au-delà des

pass, it was still sent to the tariff committee, which followed Vinchent and Curchod's directions and reworded it in order to win the approval of the assembly[76].

Differentiating between extra-European and European tariffs and working towards charging per word must be seen within a wider policy to lower telegram prices. It is revealed explicitly in the communication of the results of the tariff committee and the ensuing debate. Meydam, the committee chairman, declared that they had not been able to draw up a

---

limites de l'Europe, demeure soumise à tous les principes posés par la Convention en matière de tarifs. […] il résulte de là que, pour profiter de la faculté que lui confère la vote relatif au dernier paragraphe de l'article 32, tout Office, d'État ou de Compagnie, est tenu d'observer préalablement les deux conditions suivantes : ne proposer l'application de la mesure que pour le parcours extra-européen et obtenir le consentement unanime de tous les Offices européens et extra-européens dont les dépêches doivent emprunter les lignes. […] M. Vinchent maintient son interprétation comme étant la seule conforme aux principes qui ont toujours été suivis pour l'établissement des tarifs. Il existe un tableau des taxes fixées à Vienne. Ces taxes peuvent être modifiées par tous les États contractants réunis en Conférences ; elles ne peuvent être réduites, en dehors des Conférences, que par le consentement de tous les États intéressés et dans les conditions prévues par la Convention. […] M. le Colonel Meydam demande que l'interprétation résultant des observations de M. le délégué de la Belgique se fasse jour dans la rédaction de dernier paragraphe de l'article 32, que la Commission spéciale a mission de formuler. […] La Conférence admet sans objection que, suivant l'interprétation de M. Vinchent, le vote de la séance précédente doit s'entendre dans ce sens que l'admission facultative de la dépêche de dix mots et de la progression par mot ne peut avoir lieu que sous le condition du consentement unanime de tous les États situés sur le parcours des dépêches auxquelles ce mode de taxation devrait s'appliquer. La Commission de rédaction est invitée à préciser cette pensée dans le texte qu'elle aura à soumettre à la Conférence à ce sujet», ibid., 380-383.

[76] «M. Curchod explique que la Commission a été amenée par des difficultés de rédaction à déterminer la quotité des réductions applicables aux dépêches de dix mots et à la progression par mot. Pour donner satisfaction aux demandes qui lui sont adressées, il proposerait de substituer au projet de la Commission les dispositions suivantes : "Toutefois les Offices extra-européens sont autorisés à admettre sur leurs lignes la dépêche de dix mots avec taxe réduite, ainsi qu'à employer la gradation par mot, après avoir obtenu le consentement des autres Offices intéressés, conformément aux dispositions de l'article 34. Pour les parcours européens, cette dépêche est taxée conformément aux dispositions du paragraphe précédent". Cette rédaction étant considérée comme satisfaisante par les différents membres qui avaient présenté des objections contre le projet primitif, la conférence décide successivement que la faculté de la dépêche de dix mots et celle de la gradation par mot seront maintenus en faveur des Offices extra-européens, et elle adopte la nouvelle rédaction proposée par M. Curchod», ibid. 578.

written report because of the open disagreement over some unresolved issues:

1° The Austro-Germanic Telegraph Union having ceased to exist since 31 December 1871, there are grounds for identifying the charges once common to the states making it up …
2° With Great Britain being admitted to the agreement between the states, the form with which the charges of these countries appeared before the Convention have had to be modified …
3° British (British India) terminal tariffs have been increased;
4° Tariffs have been introduced for the Dutch Indies and Indochina and tariffs for Algeria and Tunisia differentiated from the French ones;
5° Lastly, Turkish tariffs have been increased because of using lines not foreseen on the old tables[77].

A part from the question of the new German tariffs, which was resolved after animated repartees between Vinchent and Meydam [78], with

---

[77]«1° L'Union télégraphique austro germanique ayant cessé d'exister depuis le 31 décembre 1871, il y a lieu de distinguer les taxes autrefois communes des États qui la composaient […] ; 2° La Grande Bretagne ayant été admise dans le concert des autres États, l'on a dû modifier la forme sous laquelle les taxes de ce pays figuraient auparavant dans la Convention […] ; 3° Les taxe terminales de la Grande Bretagne (Inde britannique) sont augmentées . 4° L'on a introduit les taxes des Indes néerlandaises et de la Cochinchine et distinguée la taxe de l'Algérie et de la Tunisie de celle de la France; 5° Enfin, quelques additions ont été apportées aux tarifs ottomans, en raison du parcours de lignes non prévues sur les tableaux anciens», ibid., 613.

[78] Here below the most significant passages of the exchange: «M. Vinchent a été heureux d'entendre M. le délégué de l'Allemagne constater la nécessité pour les Communications à petite distance de laisser les tarifs à la portée de tous. Cela prouve qu'il reconnaît qu'en pareil cas le relèvement des taxes est un mal. Tel est cependant le résultat auquel on arrive à la suite de la dissolution de l'Union austro-germanique. Cette Union qui a si longtemps contribué au progrès de la télégraphie était parvenue à donner au public la taxe de 3 francs. Elle cesse d'exister pour des motifs que la Conférence n'a pas à examiner ; mais le public doit-il en souffrir ? La Belgique, notamment, se trouverait atteinte fortement par les nouveaux tarifs que réclament les anciens États de l'Union. Au lieu de 4 francs, la taxe d'une dépêche de Bruxelles pour Vienne serait de 7 francs. C'est le premier exemple d'une augmentation des taxes pour des communications à petite distance et toute l'attention de la Conférence doit être attirée sur des conséquences aussi anormales. […]M. Vinchent a consulté son Gouvernement et, d'accord avec les instructions qu'il a reçues, il ne croirait pas pouvoir signer un acte qui consacrerait une disposition aussi défavorable aux intérêts du public belge», ibid., 626. Meydam answered as follows : «Il ne croit pas que le public belge puisse être considéré comme le seul juge des travaux de la Conférence. Le but qu'elle veut atteindre est

the latter's promise to lower charges in Germany[79], most of the debate was taken up with telegram tariffs for India.

The Vienna Conference had fixed some charges, the sum of which was the total price of a telegram sent to India in a moment in which Great Britain, which had most stakes in the flow of telegrams towards Asia, was not yet in the TU, and neither were most private companies present. Later, in September 1871, a special committee created in Vienna had its one and only meeting in Berne and decided on the increase of extra-European tariffs to India. Not all TU members recognized the new tariffs, like Turkey, which was directly involved in the question, though it had not

moins exclusif», ibid., 629. The spat went on in the next sitting, bolstered by Ponzio Vaglia, too : «M. Ponzio Vaglia n'est pas surpris que M. le délégué de la Belgique ait invoqué les intérêts du public belge. Tous, sans doute, sont persuadés que les effets de l'augmentation seront mauvais partout, mais chacun des délégués parle, avec meilleure connaissance de cause, des intérêts qui touchent à ses administrés», ibid., 632 ; «Pour répondre à un fait personnel, il [Vinchent] demande à ajouter quelques mots. Il a regretté d'entendre, dans la dernière séance, M. le délégué de l'Allemagne prononcer le mot de reproche. Les relations que M. Vinchent a entretenues pendant de longues années avec M. le Général de Chauvin et l'appui qu'il a toujours trouvé en lui pour la propagation des idées les plus libérales, sont une garantie de ses dispositions à l'égard de l'Administration allemande», ibid., 636; «M. le délégué de l'Allemagne ne veut pas que l'on considère le public belge comme le seul juge de la question ; M. Vinchent n'a jamais eu une pareille prétention. Si l'on consulte la statistique, l'on reconnaîtra que le public belge n'occupe pas la dernière place ; mais lorsque le délégué de ce pays a parlé des correspondances de Bruxelles avec Vienne il a simplement voulu citer un exemple qui intéresse, d'ailleurs, autant le public de Vienne que celui de Bruxelles», ibid., 637 ; «Meydam désire faire une observation personnelle. Il n'a jamais voulu mêler les considérations de personnes qux questions en discussion. Il a pu se trouver parfois en désaccord d'idées avec M. le Délégué de la Belgique, mais ces divergences d'opinion de sauraient influer sur ses relations avec lui ni sur l'estime toute particulière qu'il a pour l'un des représentant les plus anciens et les plus éclairés de la télégraphie internationale», ibid.

[79] «Pour donner satisfaction aux demandes réitérées qui se sont produites dans le but d'atténuer les conséquences de la dissolution de l'Union austro-germanique, M. le Président fait connaître que l'Allemagne et l'Autriche-Hongrie fixent réciproquement à 2 francs leurs taxes terminales et de transit pour toutes les correspondances européennes qui empruntent le réseau de ces deux États. M. le Colonel Meydam est tout disposé à examiner, en outre, les concessions qui lui sont demandées et à en tenir compte, autant que la chose lui sera possible ; mais, avant de faire connaître des chiffres définitifs, il doit s'entendre avec les représentants des différents États qu'affecte la nouvelle situation de l'Allemagne. M. Vinchent rend hommage aux sentiments de conciliation que manifeste M. le délégué de l'Allemagne.», ibid., 657.

taken part in the Berne meeting. And what happened was that by refusing to raise its tariffs Turkey brought back into operation the regulation for the cheapest routing and so forced most offices to send their telegrams through Turkey.

Given that the Rome-approved non-competition pact forbidding both reductions and increases along the same route was not yet valid, Turkey was asked to accept this increase. The question was extremely sensitive, both from a formal and substantial point of view[80]. Formally the delegates had to decide what kind of procedure to adopt and since in theory the tariff tables had to be approved unanimously they were not subject to the simple majority rule adopted for all the other convention and regulation norms[81]. Consequently, in the case of disaccord three decisions could be taken: 1) they could operate a majority vote for the tariff table, running the risk of a veto from the opposition; 2) they could leave the single states totally free from multilateral agreements and face probable international increases; 3) they could seek at any cost an agreement between the administrations present and presumably also with the private companies[82].

Substantially, the liberal ideology initially adopted in Paris came strongly to the fore in the progressive reduction of international telegrams. Not all states were however in agreement, especially those like Great

---

[80] «M. Vinchent pense que la discussion doit aboutir à un résultat, et ce résultat sera l'expression de l'opinion de la Conférence sur certaines questions, telles que la légitimité de l'augmentation des taxes indiennes ou l'avantage d'égaliser les tarifs de toutes les voies», ibid., 641.

[81] The Italian delegate recalled the rule: «M. Ponzio Vaglia croit que l'on peut voter un vœu. Les questions conventionnelles se décident à la majorité des voix, sauf le droit de veto de chaque État. Quant aux tarifs, il faut l'accord unanime des États intéressés», ibid.

[82] «La conférence ne pouvant pas imposer les tableaux des tarifs, M. le Président pense que la fixation doit être établie seulement par les États intéressés dans le parcours. Quant aux autres Administrations, elles accepteront ce que ces États auront admis. Mais, si l'accord ne s'établit pas, le résultat que la voie cessera d'exister pour le service international. Il demande donc à l'Assemblée si elle croit possible de voter autre chose que l'expression d'une opinion», ibid., 646 ; «M. Vinchent ne serait pas en état de répondre à la question posée par Meydam. Il lui paraît très désirable d'avoir des tarifs égaux, mais ce serait payer cet avantage bien cher que de l'acheter par un relèvement de taxes et ce serait le payer bien plus cher encore, si ce résultat ne pouvait être atteint que par une pression exercée sur un État pour obtenir une augmentation forcée. La question est donc difficile à résoudre. Si l'accord pouvait s'établir sur la base d'une réduction, d'un côté, et d'une légère augmentation, de l'autre, la solution serait préférable. Mais cet accord est-il possible?», ibid., 647.

Britain, which wanted to keep international prices high in order to cover domestic losses[83]. In reality, other European states covertly carried out the same policy while calling for a reduction of charges[84]. In European countries with high levels of transit (e.g. Italy, Belgium and Switzerland) revenue from international telegrams played a big part in the overall income of the administrations[85]. As they were small in dimension with revenue mainly from transit charges, they preferred to contain prices in

[83] This policy was spelled out by Robinson, the British India delegate: «Si le recettes arrivaient à couvrir les frais, le Gouvernement indien serait heureux de diminuer les tarifs; mais, dans les conditions actuelles, il trouve injuste de faire supporter à la masse des contribuables les frais d'une exploitation qui n'est utilisée que par de riches commerçants, en étant de payer le prix du service qu'ils demandent», ibid., 638.

[84] Curchod, for example, preached lowering tariffs and held that international tariffs could not compensate for domestic expenses : «Il [M. Curchod] demande si le réseau indien a été construit exclusivement pour le service international et si sa principale raison d'être n'a pas été les exigences de l'administration intérieure de ce vaste pays. Quelques taxes que l'on impose aux correspondances internationales, elles seront toujours insuffisantes à couvrir les frais d'une Administration aussi considérable. Ce n'est donc pas là que l'on doit chercher la rémunération des dépenses effectuées. Quant à l'intérêt du public, au nom de la Suisse, qui a avec les Indes un trafic qui n'est pas sans importance, M. Curchod peut affirmer que les augmentations déjà introduites à Berne n'ont pas été accueillies favorablement. Les réclamations seraient bien plus fortes, si les taxes subissaient un nouvel accroissement. D'Ailleurs, les autres États pourraient ne pas admettre ces relèvements de tarifs sans y répondre par une élévation de leur propre taxes et ce serait alors une augmentation générale indéfinie. M. Curchod reconnaît que de semblables questions ne peuvent être résolues par une majorité, mais il espère qu'en voyant l'indien consentira à restreindre ses tarifs aux chiffres adoptés par la Commission de Berne», ibid, 642.

[85]In his article quoted above, Vinchent explains this concept very well: «La correspondance à l'intérieur de chaque État se fait à perte, le télégramme interne ayant le minimum de taxe avec le maximum de prix de revient. La correspondance internationale produit de bénéfices. Mais, comme la première est plus susceptible que l'autre de se développer d'année en année ; comme d'ailleurs la dépense augmente graduellement par l'élévation du prix de toutes choses et par l'obligation constante d'améliorer le service, la somme des dépenses n'est plus couverte par la somme des recettes. Quel est le remède que comporte cet état de choses ? Les Gouvernement peuvent-ils ou doivent-ils l'appliquer?», Vinchent, *Considérations générales*, 328. A further and more valid proof of the weighty contribution of international traffic is to be found in : *Statistique générale de la télégraphie dans les différents pays de l'ancien continent* (Berne: Bureau International des Administrations Télégraphiques-Imprimerie Rieder & Simmen, 1871).

order to increase traffic and income[86]. Differently, the metropolitan British and Indian administrations were the two extreme terminals of the European network. They were either the starting or ending points of telegrams and consequently had little truck with transit operations, at least until 1866 and the laying of the transatlantic cable. Given their different locations within the European network, for Great Britain, India and in a certain sense Russia, it was to their advantage to increase destination charges, rather than stimulate traffic via a generalized reduction which would help states with much transit traffic. In sum, there were two different network philosophies in the conference. On one side the peripheral countries on the edges of the international system which were in favour of tariffs favouring end-to-end flows, on the other the central European states which preferred a centripetal network privileging transit in its core area.

The debate soon turned into a stand-off, with everybody including the private companies against the increase of Indian charges[87]. Once the

[86] Not everyone agreed over this policy. A possible reference was made directly by Curchod at the Vienna Conference, when differently from above he made a communication as Head of Swiss Telegraphs. «Si le système de l'amendement eut prévalu en 1865, la Suisse aurait dû renoncer à toute part dans le transit entre l'Europe occidentale et l'Europe Orientale; car ce n'est que par des réductions sur les taxes de la Convention, opérées de concert avec l'Autriche, qu'elle a pu attirer sur ses lignes, soit les correspondances échangées entre la France et l'Autriche, soit celles de même origine pour la Turquie, la Russie, etc. L'office Suisse n'a pas seul profité de ces combinaisons ; elles ont encore tourné à l'avantage du public, en ouvrant à ses correspondances des voies multiples. M. Curchod partage toutefois l'avis général que c'est par la nature du service et non par des réductions de taxes exagérées que les États doivent lutter entre eux, et, tout en maintenant le principe de liberté, il ne refuserait pas d'y introduire une restriction dans ce sens», *Documents de la Conférence Télégraphique Internationale de Vienne*, 308.

[87] «M. William Siemens fait connaître que la Compagnie indo-européenne et, sans doute aussi, les Compagnies sous-marines, en préparant leurs tarifs, avaient pris pour point de départ le maintien des taxes de Berne. Il ne s'attendait pas à l'augmentation demandée par le Gouvernement indien et il voudrait que cette question fût vidée avant que l'on discutât les taxes de transit», ibid., 643 ; «La clientèle des Compagnies sous-marines unies représentant plus de la moitié des personnes qui correspondent avec les Indes, M. Despecher croit pouvoir prendre la parole pour le public anglais. Ces Compagnies repoussent l'augmentation demandée par le Gouvernement indien qu'elles trouvent exorbitante. […] une pareille alimentation pour une ligne qui a coûté un million est plus que suffisante pour donner de beaux bénéfices et, si le Gouvernement indien trouve l'exploitation de cette ligne trop onéreuse, les Compagnies sont toutes disposées à la prendre pour leur compte», ibid., 645.

prevailing mood had been assessed[88], the chairman of the sitting, who was not in a position to reject the decision of a single state, suggested voting in favour of a common note declaring that the high Indian charges were not justified[89]. As could be imagined, the note was approved, much to the annoyance of the Indian delegates, who later put out a communiqué in which they declared their desire to go on increasing charges[90]. Finally, thanks mostly to the Turkish mediation[91], the charges for India were slightly lowered[92]. In spite of this, Turkey was left with the lowest charges and the "Turkish route" towards India continued to be the first choice[93]. Yet again, the conservative no-policy stance of the founding states won over any proposals for innovation.

---

[88] «M. le Président s'adresse aux représentants de la Grande Bretagne et des Indes. Toute la Conférence insiste pour que l'Inde renonce à son tarif qui est jugé trop élevé. Le public intéressé avant tout au bon marché est le public anglais. Est-ce qu'un nom de ce public, personne ne proteste contre une aussi forte augmentation ?», ibid., 643.

[89] «M. le Président constate que la Conférence ne peut voter sur une taxe proposée par un Etat, mais elle peut émettre une opinion. Il serait d'avis de soumettre la question suivante à la votation : "la taxe terminale proposée par le Gouvernement indien est elle justifiée ?"», ibid., 645.

[90] «Avant que la discussion soit ouverte sur cette nouvelle proposition, M. le Colonel Robinson tient à faire, au sujet du vote émis dans la dernière séance, la déclaration suivante dont il demande l'insertion au procès-verbal : "[…] Je suis tous disposé à recommander toute taxe proportionnelle que la Conférence, après un sérieux examen, aura trouvée juste, mais je ne vois aucun avantage à admettre les propositions telles qu'elles se sont actuellement produites". En ce qui le concerne, M. Robinson ajoute qu'il serait préférable de déterminer maintenant les taxes intermédiaires, pour qu'il puisse apprécier quelle taxe il doit y ajouter», ibid., 653.

[91] «Yanco Effendi croit que l'on touche à une entente et il espère que MM. Les délégués des Indes se prêteront à cette conciliation. […] En résumé, tout en ne repoussant pas d'une manière absolue le chiffre de 94 francs, Yanco Effendi prie la Conférence de l'abaisser à 90», ibid., 655-656.

[92] «M. Robinson déclare que, tout en continuant à ne pas trouver équitable la répartition proposée en ce qui concerne la taxe indienne, il est prêt, dans une pensée de conciliation, à l'accepter et à la recommander à l'approbation de son Gouvernement», ibid., 656.

[93] «Il résulte de cette disposition que les dépêches du continent, comme celles de l'Angleterre, seront moins élevées par la voie de la Turquie que par les autres lignes», ibid., 664.

5.2.3 New projects, old solutions

At St Petersburg (1875) the tariff system for European telegrams seemed more anachronistic than ever and in sore need of change. Two innovative proposals were put on the table, the first by the Austrians at the beginning of the debate on tariffs, the second by the Germans as an act of provocation during the closing debate of the conference. There was also a third unusually innovative proposal from the tariffs committee, which suggested extending the extra-European regime of a simple ten-word telegram and charging any extra by the word. Yet despite all the contradictions, defects and anachronisms in the European charge system, no changes were made, no one was bold enough to insist.

The Austrian Brunner was the first to present a tariff project which aimed at withdrawing any discretionary power over transit charges. They were to be fixed according to an assembly-approved coefficient which calculated only distance[94]. Brunner clearly wanted to do away with the

---

[94] «Il sera établi, pour toutes les dépêches échangées entre les bureaux de deux États contractants qui ne sont pas limitrophes et qui, par suite, empruntent le transit d'un autre État, une taxe uniforme, quelle que soit la voie de transmission. Toutefois, pour l'établissement de la taxe uniforme, le même État pourra être divisé en deux ou plusieurs sections géographiques. La taxe pour la correspondance entre deux États sera composée selon les principes des articles 10 de la Convention et XV du Règlement de la manière suivante : 1° de la taxe terminale de l'États de départ et de la taxe terminale de l'État de destination; 2° d'une taxe uniforme de transit établie en raison de la distance des deux États. Les taxes terminales sont fixées par les États extrêmes et peuvent, à toute époque, être modifiées d'un commun accord entre ces deux États. La fixation et la modification des taxes de transit ne peuvent avoir lieu qu'en suite d'une décision de tous les États contractants. Le décompte des taxes terminales, ainsi que celui des réponses payées et autres frais accessoires, aura lieu entre les deux États terminaux par l'intermédiaire du Bureau international. A cet effet, les Administrations établiront des registres comprenant le nombre des dépêches de départ et d'arrivée, classées d'après chaque État de destination et d'origine et réduites en dépêches simples. Ce nombre, multiplié par la taxe moyenne établie d'après l'article LXXIII du Règlement international, donnera le doit et l'avoir des États intéressés. Les taxes de transit perçues par les différents États formeront un revenu commun pour tous les États contractants et seront réparties en proportion du nombre des dépêches de transit que chaque État aura transmises et d'un coefficient à fixer pour chaque État. Ce coefficient s'obtiendra en divisant le total des recettes perçues par chaque État pour la correspondance internationale de transit, pendant une certaine période à fixer d'un commun accord, par le nombre des dépêches de transit expédiées pendant la même période. Le Bureau international sera chargé de l'établissement des comptes de la correspondance de transit et de la répartition de ses produits. A

ruling made in Paris and eliminate any advantages enjoyed by cheaper routes[95].

What happened was that despite expressing great admiration for the proposal, the delegates did not back it. The Spaniard De Tornos, for example, chairman of the tariffs committee, pointed out that though the proposal was based on a serious and useful piece of work, the times were not ripe for it to be implemented. He felt that if the Austrian proposal was introduced at once, it would lead to excessive bookkeeping complications in the short term. He thought it better therefore to put the project in the hands of the Bureau and have its feasibility and possible scheduling verified[96]. Vinchent complimented the Austrian project for being thorough and worthy of attention, but felt it was not feasible and needed passing over to the Bureau[97]. Differently from his colleagues, however, Brunner felt his tariff proposal was revolutionary:

---

cet effet, chaque État lui fournira un relevé sommaire», *Documents de la Conférence Télégraphique Internationale de St Petersburg*, 130.

[95] «En première ligne, elle [la proposition austro-hongroise] met fin à toutes les concurrences fâcheuses qui ont pour but et pour effet de faire dévier les dépêches de leur voie naturelle, en affectant aux voies détournées des taxes réduites. Si la proposition austro-hongroise est adoptée, les dépêches ne s'écarteront plus de la voie naturelle que dans les cas d'interruption», ibid., 341.

[96] «M. de Tornos [...] fait connaître que cet amendement a été également discuté par la Commission et qu'il en est résulté l'opinion que l'on devait savoir gré à l'administration austro-hongroise d'avoir produit une proposition qui constituait un travail sérieux et utile et montrait un louable désir, auquel s'associait la Commission, de simplifier les tarifs. Mais la Commission, d'un autre côté, ne croit pas la question suffisamment mûre ; elle pense qu'elle doit être étudiée et que toutes les administrations ne seraient peut-être pas en mesure d'appliquer déjà ce nouveau système. La Commission croit aussi que, telle qu'elle est proposée, la réforme porterait, peut-être, préjudice au transit de certains États, que les décomptes entre les administrations terminales seraient plus compliqués qu'aujourd'hui et que ceux du bureau ou de la caisse centrale seraient nécessairement presque toujours en retard. La Commission estime donc que l'administration austro-hongroise a rendu un service à la télégraphie internationale en prenant l'initiative d'une proposition qui, certainement, portera ses fruits ; mais elle ne croit pas que le moment de l'appliquer soit encore arrivé et elle propose de charger le Bureau international d'étudier et de préparer, d'accord avec l'administration austro-hongroise, un projet de tarifs et de décomptes internationaux qui puisse être présenté aux prochaines Conférences, en l'autorisant à réclamer directement des administrations les données nécessaires à ce travail», ibid.

[97] «M. Vinchent a combattu la proposition austro-hongroise dans la Commission, mais il la croit digne de la plus sérieuse attention [...]. Il admet donc que la

> Finally, universal history teaches us that in everything innovations meet real difficulties and M. Brunner is far from denying it in the question that he has the honour of submitting to the Conference. But history itself shows that all innovations made in the spirit of historical development are accompanied by a real success and furthermore, though it can be delayed, it always ends by coming through [...] If the members of the Conference adopt this proposition, the consolidation of the Telegraph Union is achieved. If they reject it, it is something for those who come after them to do[98].

Unsurprisingly, the assembly was not of the same opinion, so the project was passed over to the Bureau for a detailed examination of its potential effects. What was clear is that the project could never have been entertained because of its effects on national sovereignty - its underlying principle of a collegially decided transit charge was on a collision course with the principles of liberty as established at Paris[99].

The delegates showed their disapproval of current charges when the tariff committee suggested extending the extra-European norms to include Europe[100]. The defenders of the present system interpreted their suggestion

---

question soit soumise à une étude préalable par le Bureau international, auquel on attribuerait les fonds nécessaires ; mais il voudrait que les auteurs de l'amendement donnassent un programme précis, avec chiffres à l'appui, et il demande que muni de ce programme le Bureau international établisse, pour une période de 1 ou de 3 mois, un décompte fictif avec le tarif actuel et avec le tarif à proposer, ainsi que le relevé des frais et du temps nécessaires pour l'établissement et la liquidation des comptes avec la caisse commune», ibid., 343.

[98] «Enfin, l'histoire universelle apprend qu'en toutes choses les innovations rencontrent des difficultés réelles et M. Brunner est loin d'en nier l'existence dans la question qu'il a l'honneur de soumettre à la Conférence. Mais cette même histoire montre que toute innovation faite dans l'esprit du développement historique est accompagnée d'un succès réel et, en outre, que si elle peut être retardée, elle finit *toujours* par percer. [...]Si les membres de la Conférence adoptent cette proposition, la consolidation de l'Union télégraphique est une chose *faite*. S'ils la repoussent aujourd'hui, c'est une chose *à faire* par ceux qui viendront après eux», ibid., 342.

[99] «Il [M. Vinchent] termine en faisant observer qu'a son avis une Union entre États, pour se maintenir toujours cordiale, e doit pas créer une communauté trop intime, qui risquerait de faire naître des froissements réciproques», ibid., 344.

[100] «La majorité de la Commission a donc l'honneur de proposer à la Conférence d'adopter le paragraphe 1 de l'article XV dans la teneur suivante: 1. Le minimum de la taxe s'applique au télégramme dont la longueur ne dépasse pas dix mots. La taxe applicable au télégramme de dix mots s'accroît de .... par chaque mot au-dessus de dix. [...] 2. Pour la correspondance extra-européenne est adopté le tarif par mot pour tout le parcours», ibid., 373.

as a challenge to the traditional principles of the TU, and Vinchent went as far as to ask for the names of the committee members who had dared to vote in favour of a change in charges[101]. Frey went on to insinuate that the new system would certainly cause a drop in income and increase in costs for all international telegrams[102]. An ugly forecast for all those administrations in Continental Europe who managed to balance their books thanks to income from international traffic[103]. Frey, Vinchent and Chambre held the same position, that lowering international and domestic tariffs followed totally different dynamics[104]. This group felt that while a

---

[101] «Il [M. Vinchent] aurait voulu, [...], connaître les noms des votants, afin d'apprécier l'influence que les intérêts de chaque Office pourraient avoir eue sur la détermination de leur représentant, et les taux des tarifs pour se rendre compte des conséquences pratiques de la mise en application du principe», ibid., 377.

[102] «M. Frey se prononce également pour le maintien du système actuel et c'est dans ce sens qu'il a voté au sein de la Commission. La raison principale est qu'il redoute les conséquences financières de toute réduction de taxe. Il admet que la réforme opérée par l'Italie dans son service intérieur ait donné des résultats avantageux et il en trouve la confirmation dans l'exemple de la Suisse et de la Belgique, où la réduction de 1 franc à 50 centimes de la taxe intérieure, a été suivie immédiatement d'une augmentation du produit brut. Mais la Suisse a fait une autre expérience. En 1872, elle a réduit de 2 fr. à 1 fr. le prix de ses dépêches avec l'Allemagne du Sud et cette réduction n'a pu, après 3 ans, qu'augmenter de 30 pour cent le nombre des télégrammes en sorte qu'elle s'est traduite en résumé par une diminution sensible de recettes, une augmentation notable de frais, et un développement de trafic bien peu important, si l'on tient compte de l'accroissement normal des relations. La même échelle ne saurait donc s'appliquer à la correspondance intérieure et à la correspondance internationale. Dans la première, toute réduction surexcite les intérêts de famille qui sont trés sensibles à une taxe plus modérée ; dans la seconde, il ne continue à s`échanger que des dépêches commerciales, indifférentes au prix.», ibid., 374-375.

[103] See again: *Statistique générale de la télégraphie.*

[104] «En résumant, sous forme de principes. Les résultats de l'expérience, tels qu'ils viennent d`être analysés, on peut considérer comme démontré: 1 Qu'une réduction du tarif télégraphique amène une diminution de produit net même dans les conditions les plus avantageuses qui nous soient connues ; 2 Que le service intérieur, déjà en déficit avant les dernières réductions de tarif, a obtenu, par ces mesures, un énorme accroissement de correspondance avec un légère augmentation de ce déficit; 3 que le service international, qui a toujours donné un bénéfice, a trouvé, dans des mesures analogues, un développement de correspondances beaucoup moindre et une diminution de ce bénéfice beaucoup plus considérable que l'accroissement de déficit intérieur; 4 qu'à l'intérieur, le déficit tend à diminuer, tandis qu'avec l'étranger la diminution du bénéfice tend à augmenter. Les motifs d'intérêt public qui déterminent une réduction du tarif intérieur ne sont donc point généralement applicables au service international», Vinchent,

radical reduction of domestic tariffs would encourage a greater use of the service, the same could not happen with international telegrams. The domestic client could be any citizen and therefore a reduction in telegrams prices would encourage private telegrams as well as business and financial users. Diversely, international users were mainly the press and the business world, who would anyway send a certain number of telegrams regardless of price variations, given that sending telegrams was essential to good business results[105]. This line of thought, often found in Belgian

---

Considérations générales, 350; «A statement of the motives for maintaining a higher tariff for international telegrams demonstrates the following principles ; 1st That in Belgium, notwithstanding the existence of the most favourable circumstances, *all reduction of tariff has resulted in a diminution of the net product.* 2nd That in the interior service the diminution of net product has been of small importance, and has been accompanied by an immense development of correspondence, that is to say, *of service rendered to the public.* 3rd That in the international relations *the reduction of the tariff has resulted in a considerable loss, with a much less development of correspondence,* as compared with the internal service, that is to say, *less service rendered to the public.* Thus the Government is justified before all in reducing the tariff upon inland telegrams, and in deferring the reduction of the international tariff until the increase of traffic overtakes the deficit in revenue caused by the reduction», Anderson, *Statistics of Telegraphy*, 279-280.

[105] For Vinchent: «Les administrations télégraphiques sont presque toutes dans cette situation de subir pour la correspondance intérieure des déficits qu'elles compensent, en tout ou en partie, par les bénéfices de la correspondance internationale ou de transit, tandis que le prix de revient de la dépêche intérieure, au contraire, par cela même qu'elle nécessite plus d'opérations, est plus fort que celui de la dépêche internationale, lequel dépasse à son tour celui de la dépêche de transit. Cette anomalie se justifie, aux yeux de M. Vinchent, par la masse des intérêts de famille que favorise, dans le service intérieur, toute réduction de taxe, et qui n'existent pas pour le service international où les dépêches presque exclusivement commerciales restent toujours à peu près les mêmes, non pas que l'expéditeur soit insensible à leur prix, mais parce qu'il est quelque sorte, toujours obligé de les expédier. Les résultats d'un abaissement dans le service intérieur et dans le service international sont donc tout différents et c'est ce que M. le ministre des travaux publics de Belgique a fait ressortir, en 1870, à la suite d'une réclamation qui s'était produite à la Chambre des Représentants, à Bruxelles, en vue d'une réduction du tarif de ce dernier service», *Documents de la Conférence Télégraphique Internationale de St Petersburg*, 376-377. For Chambre : «M. Chambre est au nombre de ceux qui ne peuvent appuyer, pour les dépêches internationales, la réduction de la dépêche simple de vingt à dix mots, avec gradation par mot ou par série de cinq mots. Il ne désire nullement prolonger une discussion déjà très-élucidée par les développements exposés dans un sens ou dans un autre, mais seulement expliquer sommairement les motifs qui déterminent les

and British publications[106], was based on concrete cases of the application of both domestic and international tariff policies[107]. However, others were

---

votes des délégués de la Grande Bretagne. Ils peuvent invoquer les mêmes résultats. Ces statistiques prouvent que le minimum de dix mots à moitié prix produirait une diminution sérieuse de recettes que l'Administration britannique ne serait d'élever dans une mesure considérable le prix des dépêches de vingt mots qui continueront à entrer dans une forte proportion dans les communications du commerce et de la presse. Cette dernière faisant, en Angleterre, comme on sait, un large usage de la télégraphie internationale, l'augmentation du prix de ses correspondances donnerait lieu, de la part du public et des journaux, à des plaintes très sérieuses qui, dans l'opinion des délégués britanniques seraient des plaintes très fondées», ibid., 384.

[106] Vinchent had thus detailed the concept in the article published in JT : «les lignes télégraphiques son occupées, pour les neuf dixièmes, par deux courants principaux : les affaires commerciales, qui utilisent le télégraphe avec plus ou moins d'empressement, d'après les circonstances politiques ou autres qui activent ou ralentissent les transactions ; les relations de famille et de société qui se développent d'année en année à mesure que ce moyen de correspondance entre dans les mœurs et, par les réductions successives des tarifs, pénètre graduellement dans les régions moyennes et inférieures de la fortune privée. Ainsi qu'on doit s'y attendre, ces deux courants de correspondances ne se présentent point dans le même rapport à l'intérieur et avec l'étranger. Les liaisons privées, les relations de famille sont beaucoup moins nombreuses au-delà des frontières. Elles n'entrent qu'à raison de 28¼ % dans les correspondances internationales, tandis qu'elles constituent 59½ % du trafic intérieur. En compensation, les télégrammes du commerce absorbent 56¼ % de la correspondance internationale et les 34% seulement du service intérieur. Avant les réductions de tarif, la proportion était égale. A l'intérieur, et les correspondances commerciales formaient les 63% du mouvement international. Il est donc évident que les correspondances suivent, surtout dans leur mouvement progressif, les fluctuations des affaires et ne sont influencées que très secondairement par le tarif. Comme elles constituent l'objet principal des relations télégraphiques internationales, on ne peut, dans ces relations, obtenir d'un tarif à bon marché l'augmentation du trafic que la même mesure amènerait dans le service intérieur», Vinchent, *Considérations générales*, 349. The same concept also appears in Anderson, *Statistics of Telegraphy*. Both Vinchent and Anderson based their considerations on the official statistics for the Belgian Telegraphs, published by the Belgian Minister of Public Works.

[107] «M. Vinchent peut citer quelques exemples à l'appui de ses assertions. En 1861, une réduction de 9 à 5 francs de la taxe entre la Belgique et Londres a donné une augmentation de dépêches de 43 pour cent. Des réductions analogues entre le tarifs différentiels avec le Pays–Bas et la France ont donné, en 1862 et 1863, l'une 31, l'autre 33 pour cent d'augmentation. A l'intérieur, la réduction de 1 franc à 50 centimes, faite au mois de décembre 1875, a produit immédiatement une augmentation de 90 pour cent. Se basant sur ces enseignements les instructions de M. Vinchent lui recommandent donc de maintenir les bases actuelles et l'ensemble

of a very different opinion, in particular the Italians and the German proposers of the committee's reform project[108], both holding that a substantial reduction in telegraph tariffs would anyway stimulate growth and that over the medium-length period would contribute to increasing income[109]. Though recognizing the difference between domestic and

---

des tarifs, en s'opposant à toute aggravation et néanmoins de combattre toute mesure réduisant les produits.», ibid., 377.

108 "Le tarif par mot pour les télégrammes (traduit de l'allemand d'après une communication originale de l'Administration allemande)," *Journal Télégraphique* III, no 32 (1876) : 631-635; "Le tarif par mot pour les télégrammes (traduit de l'allemand d'après une communication originale de l'Administration allemande)," *Journal Télégraphique* III, no 33 (1876) : 651-654 ; Philip Benton, "De l'effet des réductions de tarifs sur le trafic télégraphique international. Réponse aux arguments de M. Vinchent," *Journal Télégraphique* III, no 32 (1876): 639-643 ; Philip Benton, "De l'effet des réductions de tarifs sur les recettes provenant des télégrammes échangés entre États européens," *Journal Télégraphique* III, no 35 (1876): 691-694 ; M. Scheffler, "Des principes des tarifs et de la statistique," *Journal Télégraphique* III, no 36 (1876): 715-719.

109 «M. Stephan pense que le télégraphe deviendrait accessible à un plus grand nombre par la réduction du minimum des mots et qu'il en résulterait pour le trésor public un bénéfice considérable. Les exemples donnés par M. Frey, pour le trafic entre l'Allemagne du Sud et la Suisse, et par M. Vinchent pour celui de la Belgique avec les Pays-Bas, ne paraissent pas concluants, car les réductions dont il s'agit ont été faites à une époque de stagnation générale des affaires qui se prolonge encore aujourd'hui», ibid., 382 ; «Quant à la diminution des recettes de près de dix pour cent qui, d'après les calculs de M. le délégué de la Belgique, serait la conséquence de l'adoption des propositions de la Commission, M. D'Amico pense que cette perte serait sensiblement allégée par l'augmentation du nombre des dépêches et pourrait l'être plus encore en adoptant la progression par cinq mots qu'il a proposée et qui est plus favorable aux administrations que la progression par mot. Il ne croit pas, d'ailleurs, que le système proposé soit exclusivement à l'avantage des commerçants et au détriment des autres clients du télégraphe. S'il paraît plus onéreux pour les dépêches du Gouvernement et de la presse, la chose a peu d'importance puisque, pour les premières, les taxes ne font que passer d'une caisse gouvernementale dans une autre et que les secondes, par la location de fils spéciaux, échappent, pour le plus grand nombre, à ses effets. La masse du public, au contraire, subira l'influence de toute réduction de prix qui produit toujours une augmentation de la consommation et attire un plus grand nombre de clients», ibid., 379. A last backing to the need to lower international tariffs was the statistic analysis coming from Germany: «L'on peut considérer ce résultat soit simplement comme le développement naturel, soit en partie comme une conséquence de l'idée qui a prévalu dans la formation des tarifs, de favoriser le trafic intérieur dont les taxes, malgré le prix de revient plus élevé de ces dépêches, son beaucoup plus faibles que celles du trafic international moins coûteux. Que l'une ou l'autre de ces

international clients, the Germans and Italians felt the two categories were to all intents and purposes alike. Looking at the situation from the present, we can see that both theories were partly correct and that in reality, users' reactions to a price reduction depended mostly on its coefficient. What emerges clearly was that the differing positions of national delegations resulted from their basic concept of the international telegraph service as exclusive or universal.

Worth noting the harsh criticism of the current tariffs made by those pushing for reform. According to Stephan:

> the present tariff system has the three following inconveniences : 1) It takes from the public the faculty of having brief cheap telegrams, very precious for business, very advantageous for the use of the lines, very positive for revenues. 2) It makes too great a difference in the progression of charges, in adding a surcharge of 50 cents for one extra word outside the scale. 3) Finally, it uselessly overloads the wires by inveigling the public to lengthen their telegrams to reach the twenty word minimum charge limit.[110]

Furthermore, he launched a serious attack on the reasoning behind the system:

> The German delegate considers as against the principles of free exchange those who tend to unite relations and do not allow two measures for services rendered to foreigners or nationals. Such a system would consequently lead to transit routes being abandoned in a search for less

suppositions soit fondée, il en résulte dans tous le cas qu'il devient toujours de plus difficile de continuer à trouver dans le trafic international la compensation des déficits du trafic intérieur et que dans un temps peu éloigné il deviendra tout à fait impossible d'obtenir un semblable équilibre sans imposer d'une manière réellement prohibitive le trafic international. Ce résultat ne montre-t-il pas qu'il convient dorénavant de renoncer à favoriser *artificiellement* une partie du trafic au détriment de l'autre et que l'on doit revenir au principe *naturel* de considérer avant tout les frais de revient des télégrammes comme la base de la fixation des taxes applicables à la majeure partie des correspondances ?», Scheffler, *Des principes des tarifs et de la statistique*, 719.

[110] «le système actuel de taxation a les trois inconvénients suivants : 1) Il enlève au public la faculté, très précieuse pour le commerce, très-avantageuse pour l'utilisation des lignes, très féconde pour les revenus, des petites dépêches à bon marché ; 2) Il met un trop grand écart dans la progression de taxe en surtaxant de 50 pour cent le seul mot qui dépasse les échelles de gradation ; 3) Enfin, il surcharge inutilement les fils en amenant le public à allonger ses dépêches pour atteindre la limite de 20 mots correspondant à la taxe minime», ibid., 381.

expensive ones […] M. Stephan would like to put an end to a state of affairs he feels abnormal, in adopting not the ten-word telegram which is an attenuation of it, but the immediate charge per word, with a minimum, not of words, but tariff[111].

Perhaps unaware of the irony, Stephan was accusing of illiberalism those who right from Paris had defended the tariff he was now questioning in the name of the liberal ideals of free competition[112].

No doubt because of the opposition of the three main telegraph powers (Switzerland, Belgium and Great Britain), the committee's proposal was turned down. Realizing anyway that times were changing, Stephan took advantage of the general discontent about the tariff system and presented a project advocating change[113]. But also aware that times were not ripe for a

---

[111] «M. le délégué de l'Allemagne les considère comme contraires aux principes du libre échange et à l'esprit général du siècle, qui tendent à rapprocher les relations et n'admettent pas deux mesures pour les services rendus aux étrangers ou aux nationaux. Un tel système aurait pour conséquence de faire abandonner le transit du pays qui l'adopte pour chercher d'autres débouchés moins onéreux. […] M. Stephan voudrait mettre fin à un état de choses qui lui paraît anormal en adoptant, non pas la dépêche de dix mots, qui ne constitue qu'une atténuation, mais la tarification immédiate par mot, avec un minimum, non de mots, mais de taxe», ibid., p. 382.

[112] «M. Stephan éprouve une certaine appréhension à venir combattre les idées exposées par des esprits éclairés et les hommes expérimentés qui ont parlé avant lui, mais la gravité de la question ne lui permet pas de se désintéresser dans le débat et il doit surtout répondre aux arguments qu'a présentés si lumineusement M. le délégué de la Belgique, avec toute l'autorité qui s'attache à son caractère désintéressé et a à l'importance du pays qu'il représente», ibid., 381.

[113] «Pour arriver à ce résultat, et ayant toujours en vue, bien entendu, la concurrence, les taxes, tant terminales que de transit, ont dû être réglées pour *le même service* d'une manière tout-à-fait différente, ce qui a créé de grandes complications. Ces complications iront en augmentant d'année en année avec le développement du réseau télégraphique. M. Stephan ne croit pas rencontrer de contradiction s'il prétend que déjà à présent pour un grand territoire, avec des frontières étendues et des côtes dentelées, cela dépasse les facultés de l'esprit humain de pénétrer avec la lucidité nécessaire et d'apprécier d'avance toutes les conséquences que les taxes établies arbitrairement et par pure voie de compromis dans les Conférences télégraphiques auront pour le budget de l'Administration de son pays. Ce qu'il a vu et étendu dans la commission des tarifs n'a fait que fortifier en lui cette conviction. Il surgit beaucoup de questions litigieuses, par exemple celle de la réciprocité, pour les quelles le système actuel manque de la boussole qui indique invariablement le point suivant lequel l'on peut s'orienter sur cet océan de l'échange télégraphique universel qu'il s'agit de franchir. La nécessité de changer les taxes dans les différentes combinaisons qui se présentent et par suite d'un

complete new project, he only proposed a study, which could be handed over to the Bureau for feasibility test[114]. Stephan's underlying motive was to score off the already wavering tariff system[115] and show that Germany was ready to guide the change in open competition with the historical powers of the TU.

---

amendement fait par une Administration quelconque, menace d'une manière permanente l'état des recettes, le montant actuel des taxes, et jette par là de la perturbation dans les opérations budgétaires. La combinaison actuelle est exposée à des vacillements, des oscillations dont les effets peuvent se produire bien loin et bien longtemps ; et elle fait à M. Stephan l'impression d'un corps dont le centre de gravité a été déplacé», ibid., 494-495.

114 «Le délégué de l'Allemagne a exposé au sein de la Commission que son idée était suffisamment développée dans l'imprimé dont MM. Les délégués ont déjà pris connaissance et que, d'autre part, il était persuadé qu'elle ne pourrait recevoir d'application dans la présente Conférence sans trop prolonger ses travaux. Il tient seulement à provoquer une discussion, qui fasse ressortir la nécessité d'adopter un principe unique et juste pour la fixation des taxes, tant terminales que de transit, de tous les États contractants. Il considère les bases qu'il a proposées comme étant le plus propres à établir ce principe. Il serait, cependant, disposé à en accepter d'autres, s'il en était proposé, pourvu qu'elles remplissent l'objet fondamental de la proposition. Tous les membres de la Commission ont trouvé fort recommandables les désirs exprimés par M. le délégués partageassent son désir d'atteindre le but proposé, tous ne seraient peut-être pas d'accord sur le moyens d'y arriver, et que la réforme proposée étant tout-à-fait radicale, elle exigerait des travaux préparatoires qui prolongeraient trop la durée des conférences. D'un autre côté, quelques États qui aujourd'hui ne seraient pas en mesure d'adopter le principe, faute d'en avoir étudié les conséquences le seront peut-être, plus tard, après un examen approfondi», ibid. 491-492.

115 «La première supposition celle qui imputerait à la délégation allemande l'intention d'une simple critique du système actuel. Loin d'elle est cette pensée. Dans les questions de ce caractère, la simple critique négative, M. Stephan, pendant toute sa carrière publique, l'a toujours considérée comme parfaitement oiseuse ; elle lui paraît ressembler à cet être de la mythologie qui ne possédait que la voix et qui manquait de bras. Aussi s'est–il empressé de reconnaître tout ce que le système actuel a produit de bons résultats et de rendre pleinement justice à l'ingéniosité d'esprit qui l'a crée, ainsi qu'a la sagacité et à la circonspection avec lesquelles il a été adapté aux différents combinaisons qui se sont produites dans le cours des années et qui, bien des fois, ont présenté des difficultés très-sérieuses. Mais la délégation allemande est persuadée que le système de tarification, à l'heure qu'il est, n'est plus conforme au développement actuel que, faute d'un principe régulateur, il est et sera de plus en plus, dans l'avenir, une source de contestations et de dissentiments ; enfin, qu'il entrave le progrès ; ce dont M. Stephan ne manquera pas de fournir des preuves tout-à-l'heure», ibid., 493.

His proposal was radical, close to the position of Brunner[116]. It called for a flat destination charge and the introduction of a transit charge costed according to distance. More radical still was his idea to re-establish free competition on international routes, opting for the cheapest way and abolishing the cartel agreement which established the same charge on rival lines along the same route[117]. He condensed his proposal into five essential points:

> 1. Determine destination charges according to the service given, keeping in mind the reciprocity between states of more or less the same size and the proportion of service given between big and small states; 2. Base the transit charge on an established principle, a mathematical quantity like distance; 3. Fix the total charge according to the cheapest route; 4. Send the telegrams via several routes in equal quantities, where the routes offer the same advantages of precision, speed and cheapness; 5. Recognise for the Administration of the sending country, the freedom to send telegrams

---

[116] «Partant de ces principe et ayant en vue en même temps le but d'une simplification des taxes actuellement en vigueur, l'Administration allemande se permet de présenter les propositions suivantes :
Taxes terminales : 10 centimes par mot pur les États d'un territoire étendu ; 5 centimes par mot pour les États d'un territoire plus restreint ; dans tous les deux cas avec un minimum de 1 franc pour la taxe totale à percevoir de l'expéditeur. Les rayons seraient donc abolis. Le territoire de la Russie étant placé dans des conditions spéciales, on pourrait fixer les taxes terminales de la Russie à 15 ou à 20 centimes par mot.
Taxes de transit : type uniforme appliqué à raison de la distance à parcourir dans le territoire des pays intermédiaires, dans ce sens que ce serait toujours la distance la plus rapprochée entre les deux États terminaux en cause qui servirait de base au calcul du droit de transit. Ce droit serait fixé pour chaque centaine de kilomètres à 1 centime par mot, ou si cette taxe paraissait être trop basse, au moins pour le moment actuel, à 2 centimes par mot, avec minimum d'un demi-franc par télégramme pour chacun des pays intermédiaires», ibid., 490.

[117] «Lorsque plusieurs voies de transit se présentent pour la même correspondance, il sera toujours appliqué à l'échange entier en cause la taxe de transit résultante de la route la plus courte. Si dans de certaines éventualités, il était nécessaire ou opportun de diriger les correspondances par une autre route que celle qui est la plus directe, l'augmentation du montant des droits de transit qui se produirait de ce chef devrait être supportée par l'Administration du pays d'origine, dont la taxe terminale offrirait donc, pour les case exceptionnels en question, une recette un peu moindre que celle obtenue régulièrement. Il serait peut-être utile d'insérer ici une disposition. Dans ce sens que là où plusieurs voies de transit offrent les mêmes conditions de taxe, de Vitesse et de régularité, l'Office d'origine serait tenu de diriger par les différent routes autant que possible par parties égales la correspondance produite dans son pays», ibid., 490-491.

also via other routes, not necessarily the shortest, but then reduce proportionately the destination charge for this administration»[118]

Like Brunner's proposal, this German one never took off. Though the European tariff was out of date with the new more intricate and widespread conformation of the European network and the new geopolitics of European communications featuring new powers and private multinationals it was just left as it was[119].

5.3 An exclusive or universal service?

A service is normally classified as exclusive or universal according to the number and quality of users accessing it, either potentially or in reality. Yet when the effects of the political economics of a service are analyzed as in this study, it is essential to consider how those who manage the service see it[120]. If they interpret their services as exclusive, they will carry out a series of policies from the construction of the network to the tariff system designed to reflect their users. If on the contrary the service is seen as universal, a series of strategies favouring user access will be adopted, cutting across social, geographic and economic differences.

---

[118] «1. Déterminer les taxes terminales en conformité du service rendu, en tenant compte de la réciprocité entre les États à peu près de même étendue et de la proportionnalité dans le service rendu entre les grands et les petits États; 2. Baser la taxe de transit sur un principe établi, une quantité mathématique telle que la distance; 3. Fixer la taxe totale d'après la route la moins coûteuse; 4. Acheminer les télégrammes par plusieurs routes en quantités égales, là où plusieurs routes offrent les mêmes avantages d'exactitude, de rapidité et de bon marché; 5. Reconnaître à l'Administration du pays d'origine la liberté de diriger les dépêches aussi par d'autres voies que celle qui est la plus courte, mais alors diminuer proportionnellement la taxe terminale de cette Administration», ibid., 497.

[119] «1. Le minimum de la taxe s'applique au télégramme dont la longueur ne dépasse pas vingt mots. La taxe applicable au télégramme de vingt mots s'accroît de moitié par chaque série indivisible de dix mots au-dessus de vingt. 2. Pour la correspondance extra-européenne, la taxe s'établit par mot sur tout le parcours, sans conditions de minimum pour le nombre de mots, ou avec un minimum de dix mots. Le système de taxation qu'un Office extra-européen déclarera avoir adopté, sera, d'ailleurs, appliqué indistinctement à toutes les correspondances échangées avec les Offices européens», art. XV, *Règlement de service international annexé à la Convention télégraphique internationale de St. Petersburg 1875.*

[120] For a brief review of the universal/exclusive positions see : Milton L. Mueller, *Universal Service. Competition, interconnection and monopoly in the making of the American telephone system* (Cambridge MA: MIT/AEI, 1997).

As managers of national administrations and private companies, the TU conference delegates had a precise concept of both domestic and international service, which surfaced especially in the debates on the matters of tariffs. Those who believed in a universal service wanted to lower charges in order to increase accessibility, while those in favour of exclusivity worked to keep the charge for a single telegram high, in order to satisfy the particular needs of their clients by supplying technical and administrative improvements.

Present through all the conferences, the different concepts of service emerged with force on at least two occasions. The first time was in Rome (1871) when the contrast between private companies and state administrations brought up the concept of public interest. The second occasion was four years later in St Petersburg, when the debate on the European system revealed a move towards keeping telegrams charges high precisely because of their being used by trapped customers.

At the Rome Conference, the universal vision of the service, normally associated with public administrations, was contrasted with the interests of private companies in order to illustrate the irreconcilable differences between state administrations and privates, and therefore offer undeniable proof that private companies could not be allowed into the TU[121]. However, those like Vinchent and Curchod in favour of private companies, especially submarine specialists, entering the TU, spoke of a "public interest" or "general interest" shared by all those running the same kind of business[122].

---

121 «M. Ailhaud trouve que l'importance et la prolongation de la discussion prouvent la difficulté d'accorder entre eux les intérêts différents des Administrations d'États et des sociétés privées. […] Il y a là des divergences inconciliables, parce qu'elles partent d'un point de vue opposé», *Documents de la Conférence télégraphique internationale de Rome*, 287; «M. Champain veut, en développant cet amendement, provoquer la discussion sur une question très importante. Le but de la Convention est de protéger les intérêts du public en général, mais non les intérêts exclusifs d'une fraction du public, telle que la clientèle d'une compagnie», ibid., 289.

122 «Il convient d'écarter de la discussion tous les arguments tirés de la concurrence des Compagnies qui contribuent à troubler les débats. La Conférence est au-dessus de ces rivalités et est réunie dans un but plus noble, celui de l'intérêt public. Or, peut on contester qu'il ne soit de l'intérêt général, que le public, étant en présence de taxes aussi élevées que celles des communications à grande distance, puisse obtenir, par un mode de taxation spécial, une atténuation aux charges que subissent ses correspondances ? L'on arrive, sans doute, de la sorte à formuler, à côté des prescriptions générales, un règlement accessoire ; mais il n'y aura du moins qu'un régime exceptionnel, au lieu de la multiplicité des

In reality, Vinchent and Curchod both had a concept of service that was closer to private management than many of their colleagues. They felt that telegraphs had to be run with business criteria and not the methods of state administrations and were therefore interested in the categories of users who could generate profits. Nevertheless, some who believed in a universal service like the Italians, wanted private companies to join, in the name of offering a complementary service[123].

At St Petersburg, the contrast between a universal or exclusive service underlay the opposition between backers and opponents of the committee's proposal. Though all the delegates seemed to agree that domestic services were universal and should have low charges in order to encourage access, opinions on the universal service were discordant. Vinchent and Frey, the tariff reform's main critics, felt that the international service should concentrate on professional clients (businesses, press agencies), and therefore high charges were more than

---

règlementations différentes provenant de l'indépendance des diverses Compagnies», ibid., 382; «On lui objecte que, d'un côté, il [M. Vinchent] veut faciliter les adhésions des Compagnies et que , de l'autre, il défend des dispositions qui rendent cette adhésion plus difficile. Il accepte avec expressément l'intérêt général. Il considère comme étant de l'intérêt du public d'admettre, avec des taxes élevées, la dépêche de dix mots et la gradation par mot, sur les parcours extra-européens. Il considère comme étant également de l'intérêt du public, de ne pas autoriser le relèvement des tarifs», ibid., 389 ; «L'on a reproché aux Compagnies, ou pour parler plus exactement, l'on a invoqué contre elles la préoccupation qu'elles ont des intérêts de leurs actionnaires. Les conséquences que l'on tire de cet argument ne sont pas fondées. Les Compagnies accomplissent un but d'intérêt public. Elles se sont constituées pour combler les lacunes que les Gouvernements ne voulaient pas remplir ; mais l'expérience ne leur a pas appris encore quel est le véritable tarif qui convient à la fois à leurs intérêts et à ceux du public. […] Si l'on croit désirable d'amener l'accession des Compagnies, il est donc nécessaire d'introduire dans l'article 34 une disposition qui leur laisse la faculté de faire des modifications de ce genre», ibid., 399-400.

[123] «Quant aux Compagnies, M. le Président est loin d'en être l'adversaire et il reconnait les grands services qu'elles rendent ; mais elles ont des intérêts tout différents et ne peuvent, comme les Gouvernements, écarter les considérations exclusivement financières. Or, la Conférence est réunie pour réviser la Convention dans un but d'intérêt public; elle doit donc, si elle veut que son œuvre reste simple et soit durable, n'y inscrire que les dispositions qui ont l'intérêt général pour objet et laisser la plus grande liberté aux compagnies qui ont à se préoccuper des intérêts de leurs actionnaires», ibid., 385-386; «Il a simplement constaté le fait que leurs intérêts n'étaient pas toujours d'accord avec ceux du public, En ce qui le concerne, il serait disposé à leur accorder une grande liberté er il se réserve de proposer un amendement dans ce sens sur l'article 66», ibid., 401.

justified[124]. Their reasoning, that it was irrational to lower the tariffs of an exclusive service, was the same that the British private companies had used to defend themselves from nationalization in 1869. There was indeed a sharp contradiction in the critics of the 1875 reform. Scudamore, the head of British Telegraphs, who had overseen the nationalization of the private companies to make telegraphs universal[125], saw the international

[124] For Frey's comments see note 102 above. Just as explicit Vinchent's words : «les administrations télégraphiques sont presque toutes dans cette situation de subir pour la correspondance intérieure des déficits qu'elles compensent, en tout ou en partie, par les bénéfices de la correspondance internationale. C'est le résultat naturel de ce que la taxe intérieure est plus faible que la taxe internationale ou de transit, tandis que le prix de revient de la dépêche intérieure, au contraire, par cela même qu'elle nécessite plus d'opérations, est plus fort que celui de la dépêche internationale, lequel dépasse à son tour celui de la dépêche de transit. Cette anomalie se justifie, aux yeux de M. Vinchent, par la masse des intérêts de famille que favorise, dans le service international où les dépêches presque exclusivement commerciales restent toujours à peu près les mêmes, non pas que l'expéditeur soit insensible à leur prix, mais parce qu'il est, en quelque sorte toujours obligé de les expédier. Les résultats d'un abaissement dans le service intérieur et dans le service international sont donc tout différents et c'est ce que M. le ministre des travaux publics de Belgique a fait ressortir, en 1870, à la suite d'une réclamation qui s'était produite à la Chambre des Représentants, a Bruxelles, en vue d'une réduction du tarif de ce dernier service», ibid., 376-377. Again Vinchent on the same subject: «d'après les renseignements qui lui ont été fournis sur ce dernier point, il a été question de prendre pour 10 mots les 2/3 ou, pour un mot, le quinzième de la taxe actuelle. Sur cette base et en se guidant sur les indications que se présente le trafic avec l'Amérique, où existe le tarif par mot, M. Vinchent est arrivé au résultat qu'il y aurait au minimum 42 pour cent de dépêches de 10 mots et, sans entrer dans le détail de ses évaluations, que le procès-verbal pourra reproduire en note, il trouve qu'avec la taxe de 3 fr., cent dépêches, qui auraient produit avec l'ancien système 336 francs 75 centimes, ne rendront avec le nouveau que 304 francs 40 centimes, soit une perte de 9½%, c'est-à-dire de 96.000fr. pour la Belgique, qui a un trafic international d'à peu près un million de francs par an. Cette perte, M. Vinchent n'est pas disposé à l'admettre, surtout en considérant qu'en fait les 32 expéditeurs sur cent qui, d'après ses calculs, perdront au nouveau tarif, son précisément ceux qu'il faudrait favoriser, c'est-à-dire la clientèle des Gouvernements et de la presse dont les dépêches sont longues, celle des intérêts de famille dont les correspondance accidentelles ne peuvent être condensées par l'emploi d'abréviations convenues, et que les 68 expéditeurs gagnant au nouveau système seront, au contraire, presque exclusivement ceux des dépêches commerciales qui n'ont pas besoin d'être surexcitées par une abaissement de taxe», ibid., 379.

[125] The words of Scudamore, who was responsible for the nationalization redemption plan: «The telegraph is after all but a quicker post; the telegram is nothing but a condensed letter, and, other things being equal, the nation which

service as exclusive. He was not alone in harbouring such a contradiction, for Switzerland and Belgium, who in the past had encouraged a constant lowering of tariffs, were now defending a system that held prices high. Yet it was not really a change of heart, since both countries had always spoken in favour of an exclusive international service where high tariffs could balance out the lower-priced domestic service. In the early conferences they had aimed at lowering tariffs because some countries' high charges were working against important business clients. The situation had now been overturned, but for both the Swiss and the Belgians it was imperative to protect transit charges.

In a different position, those like D'Amico and Stephan, who saw the international service as an extension of the national one and therefore to be treated as universal[126]. It followed therefore that for them lowering tariffs would be one of several measures in favour of a greater access to the service. Here the influence of the establishment of the universal Post Service in 1874, which introduced a uniform tariff via international franking. It was particularly strong in some countries like Britain and Germany, where the post and telegraphs were run by the same authorities. What happened was that some of the managers of these hybrid administrations like Stephan took part in the Post Conference in Berne in

most freely uses the post will also most freely use the telegraph», British Telecom Archives, London, *Electric and International Telegraph Company*, *Opposition to the Telegraph Bill*, TGE 1/6.

126 «M. D'Amico n'estime pas, d'ailleurs, que les administrations télégraphiques doivent chercher à devenir des services productifs, car les bienfaits que les pays retire du télégraphe sont hors de proportion avec les frais qu'il coûte. Les pertes provenant du service international ne sauraient dans aucun cas être bien sensibles. M. D'Amico ne nie pas que les réductions de taxe n'aient un effet plus considérable sur le trafic intérieur que sur le trafic international, mais il y a pourtant dans ce dernier la masse des relations du petit commerce qu'arrête un tarif élevé et qui forme une clientèle intéressante à attirer par l'offre de tarifs plus modérés», ibid., 375-376 ; «Il [M. D'Amico] ne croit pas, d'ailleurs, que le système proposé soit exclusivement à l'avantage des commerçants et au détriment des autres clients du télégraphe. S'il paraît plus onéreux pour les dépêches du Gouvernement et de la presse, la chose a peu d'importance puisque, pour les premières, les taxes ne font que passer d'une caisse gouvernementale dans une autre et que les secondes, par la location de fils spéciaux, échappent, pour le plus grand nombre, à ses effets. La masse du public, au contraire, subira l'influence de toute réduction de prix qui produit toujours une augmentation de la consommation et attire un plus grand nombre de clients», ibid., 378.

1874, which gave birth to UPU and then went on to St Petersburg, taking with them the idea of an economic, flat charge[127].

At St Petersburg the Germans and the Italians were in the front line defending the idea of a universal telegraph service on the lines of the post service[128] and they suggested a reform tariffs on the lines of the postal one approved in Berne[129]. The convergence of the two services was further encouraged by the introduction of innovations like agent delivery, inspired by the postcard and the registered telegram, evidently on the model of the registered letter[130].

---

[127] This link between the backers of a tariff reform and the Postal Union had been noted by Vinchent : «M. Vinchent n'est pas à l'abri de semblables séductions et il a été heureux récemment de s'associer aux reformes libérales qui ont inauguré la création de l'Union postale dont M. le délégué de l'Allemagne a été l'ardent promoteur. Mais, de même que après deux ou trois révisions, les congrès futurs des postes se contenteront peut-être des résultats acquis, il se demande si la Conférence télégraphique, qui se trouve dans la situation où ceux-là seront plus tard, ne doit pas se défier des séductions qui la sollicitent», ibid., 376

[128] Richard R. John, "Study on universal postal service and the postal monopoly" Appendix D. *History of Universal Service and the Postal Monopoly* (George Mason University: School of Public Policy, November 2008).

[129] «M. D'Amico ne rappellera pas les objections qui avaient été opposées à sa proposition, car la combinaison de la Russie y donne satisfaction ; elle répond également aux désirs de la Belgique, d'établir une distinction marquée entre deux catégories de dépêches, dont les unes seraient entourées de garanties beaucoup plus complètes ; elle doit satisfaire enfin les membres qui craignent de supprimer le système actuel, puisque, en le conservant, elles se bornent à lui juxtaposer une autre système, comme la carte postale s'est juxtaposée à la lettre» ; «La masse du public, au contraire, subira l'influence de toute réduction de prix qui produit toujours une augmentation de la consommation et attire un plus grand nombre de clients. Aux yeux de M. D'Amico, l'essentiel, en télégraphie comme pour la poste, est de rendre les services accessibles à très bon marché. Quant aux facilités plus grandes, attribuées par l'élévation du poids ou du nombre des mots, à la lettre ou à la dépêche simple, elles ne constituent que l'accessoire. Des objections se sont basées sur la complication de la taxation, mais elles disparaissent si l'on adopte la gradation par série au lieu de la progression par mot. M. D'Amico insiste, d'ailleurs, sur ce que l'ensemble de ses amendements introduit des restrictions nombreuses dans la réduction de la taxe simple», ibid., 378-379.

[130] «La modération de taxe amènerait toute une catégorie de consommateurs qui manque à présent, ce qui constitue, en même temps, un avantage public et une compensation pour les Administrations. Quant au dilemme de M. Brunner, M. D'Amico ne le trouve pas juste, et il en sera de même pour lui que pour celui qu'on a fait autrefois au sujet de la carte postale qui devait tuer la lettre ou ne pas vivre. Il n'est pas exact que le public se plaigne d'être moins bien servi quand on ne lui demande que peu», ibid., 522.

5.3 Conclusions

As we have shown, the TU's tariff policy was the result of a far more complex process than agreeing over the standardizing of both norms and technology.

A main difference was that the tariff norms had not been handed down from the pre-TU conventions, but were the fruit of debating and the compromises that emerged during the course of the conferences. Secondly, the fact that the tariffs were generated by discoursing and deliberating meant that there was no predefined and regular course to follow. In this way the policies they were following in theory somehow got lost upon the way. For example, in spite of all the talk about tariff reductions, international charges stayed at far higher levels than domestic charges up to WWII. Then while European tariffs seemed almost comatose, extra-European charges were rapidly simplified and considerably lowered by the charge per word system. A further complication was individual states proposed tariff strategies which were the expression or precise concepts of an international service that were not always in line with their domestic service.

The multiple compromises between state administrations and state versus private companies were the main reasons for the complexity of the tariff system and the notable contradictions it gave birth to. For example, the administrations defended the principle of free competition (the cheapest route) but at the same time created a cartel agreement (the anti-competition norm). They wanted to keep twenty word telegrams in Europe but at the same time payment by word for the extra-European area. They spoke for a gradual reduction in telegram prices but at the same time regularly approved tariff tables with prices creeping up between one conference and another.

So while the standardization of norms and technology highlights the almost monolithic inertia of the TU, its tariff policy is emblematic of its compromising nature and shows to the full the complexity of governing political and economic dynamics in a state of continuous tension.

CONCLUSION

The dichotomy "innovation/conservation" was indeed a feature of the TU right from its origins. Following a mechanism like that of first mover businesses coming into a new market with innovative technologies and organizational system, the TU set up mechanisms of self-preservation to stall possible new players or modifications of the system both on the international scene and in service management. An authentic defence of a monopolistic management of the international network, therefore, which found consensus in all the players involved.

Yet, when it was established in 1865 the TU was the first supranational government organization and descendent of the telegraphic multilateralism of the period 1848 to 1865. Thus it took over both the norms and the structure based on the plenary assemblies of diplomats. Right from the beginning consolidated customs and equilibriums, i.e. continental Eurocentrism, international liberal-flavoured politics, all went towards favouring the immediate creation of mechanisms working at its self-conservation. A first example is the progressive tendency to bureaucratize the processing and approval of norms and gradually depowering the delegates. A second is the establishment of a permanent body which besides carrying out a generic role of intermediary and supervisor ended by giving the TU itself an institutional policy. The third is the refusal to give private companies the right to vote, which segregated them off into a consultative role.

The consequences of this policy can be seen in how the TU carried out its main functions – of standardizing norms/technology and creating uniform tariffs. In all these cases a kind of conservative somnolence led to the norms of the conference only undergoing slight modifications, thus conserving the original pre-TU set-up. It is indeed surprising that this kind of semi-paralysis affected the dynamics of technical standardization, where, contrary to what much of the literature holds, there was little attempt to be on the cutting edge.

Regarding tariffs, especially for extra-European telegrams, the conservative stagnation was not so complete, and allowed changes both in the criteria used for pricing and in fixing individual tariffs. In fact, both the structural changes in the international network (new intercontinental cables) and the changes in international political equilibriums (new

European members) generated a series of conflicting interests which forced delegates into a flurry of diplomatic activity in their search for compromise. Then the laying of transatlantic cables and the construction of long mixed land and submarine telegraphic routes towards Asia highlighted two problems, how to make telegrams travelling such a long distance remunerative but not excessively expensive, and how to regulate the competition with private submarine cable companies, which in many cases had hung onto the privilege of determining their own prices.

Yet the formulation of new criteria for tariffs rotated around an original compromise which grew stronger with the passing of time, and all conference participants including the privates accepted the agreement not to create competition. It was therefore the mutual recognition of this pact that allowed delegates to modify the tariff criteria slightly more than other norms.

So contrary to what many authors uphold, the TU did not revolutionize the standardising of technical aspects, norms and tariffs. More correctly, we can speak of a strategy to conserve the original *status quo* which sought not to make waves, but to preserve preceding mechanisms. In the light of these reflections, the weight of the technical elite needs to be reviewed. Initially part of the literature had thought that as the delegates taking part in the conferences were telegraph officials, they drew up the norms of the conventions according to a criterion of technical efficiency. Here we have shown that these officials behaved mainly like diplomats, followed the guidelines of their government and defended national interests. The result is that the atmosphere of the conference, conventions and minutes was imbued more by political liberalism than scientific positivism.

As other researches have shown, the states that in some way dominated the processes of formation of the convention norms were Belgium and Switzerland, the same two which had guided the phase of multilateralism. It is surprising to note, for example, how the new entry Great Britain did not change in a radical way this equilibrium. In spite of the fact that with the 1871 Conference it became the only state to enjoy a double vote (for the British Isles and the other for India) and though it was backed by the British financed submarine cable companies, Great Britain was not able to insist on its tariff proposals, and even arrived at the point of threatening to leave the organization.

Finally, the submarine cable companies had far less influence on the TU's decisions than was initially thought. Though many of their representatives were present at the conferences in a dual role as national delegates too, it was not enough to allow their line to win through. Even after 1871 when the British delegates had obtained a strong foothold, they did not manage to make private interests count. Furthermore, though the private companies managed to obtain some rare concessions from the national delegates, they did so thanks to their recognizing the anti-competition norm on international lines.

To sum up, our meticulous reconstruction of the events of the formative years of the TU allows us to refute some common places which have guided scientific literature on the history of the International Union of Telecommunications up to now: 1) the TU was not determining in creating technical standards; 2) within the TU the officials' diplomacy dominated their technical knowhow; 3) the states that guided the establishment of the TU went on dominating decision-making; 4) though private companies were present, they only had a marginal influence over the drawing up of the norms, including the tariffs.

To conclude, the TU was principally an international political body which wielded much influence over the international market of telecommunications by means of its regulations and stranglehold over competition. For this reason though it was an exciting new subject, it lacked the excitement to make it visionary.

BIBLIOGRAPHY

Primary sources

International Telecommunications Union (ITU) Archives, Geneva.

Manuscripts

Correspondance du Bureau International des Administrations Télégraphiques, 1869-1914.

Notifications du Bureau International des Administrations Télégraphiques, 1869-1914.

Printed documents

Documents diplomatiques de la Conférence télégraphique internationale de Paris. Paris: Imprimerie Impériale, 1865.

Documents de la Conférence Télégraphique Internationale de Vienne. Vienne: Imprimerie Impériale et Royale de la court et de l'État, 1868.

Documents de la Conférence Télégraphique Internationale de Rome. Berne: Bureau International des Administrations Télégraphiques, 1872.

Documents de la Conférence Télégraphique Internationale de St Petersburg. Berne: Bureau International des Administrations Télégraphiques, 1875.

Documents de la Conférence Télégraphique Internationale de Londres. Berne: Bureau International des Administrations Télégraphiques, 1880.

Documents de la Conférence Télégraphique Internationale de Berlin. Berne: Bureau International des Administrations Télégraphiques, 1886.

Documents de la Conférence Télégraphique Internationale de Paris. Berne: Bureau International des Administrations Télégraphiques, 1891.

Documents de la Conférence Télégraphique Internationale de Budapest. Berne: Bureau International des Administrations Télégraphiques, 1897.

Documents de la Conférence Télégraphique Internationale de Londres. Berne: Bureau International des Administrations Télégraphiques, 1903.

Documents de la Conférence Télégraphique Internationale de Lisbonne. Berne: Bureau International des Administrations Télégraphiques, 1909.

Statistique générale de la télégraphie dans les différents pays de l'anciens continent. Berne : Bureau International des Administrations Télégraphiques, 1871.

Archives Fédérales, Berne
Fond E-52, Nr. 440.
Fond E-52, Nr. 441.
Fond E-52, Nr. 443.

Journals
Annales Télégraphiques, Paris, 1855-1897.
L'Année scientifique et industrielle, Paris, 1857-1908.
The Electrician, London, 1861-1935.
Journal des Télégraphes, Revue mensuelle Internationale, Paris, 1865-1868.
Revista de telégrafos, Madrid, 1868-1877.
Journal Télégraphique, Bern, 1869-1914.
The Proceedings of the Institution of Electrical Engineers, London, 1872-1914.
The Electrical Review, London, 1873-1914.
L'Electricité, Paris, 1876-1885.
L'Electricien, Paris, 1881-1914.
Il telegrafista. Periodico di telegrafia, Roma, 1881-1888.
Giornale delle comunicazioni, Roma, 1881-1891.
Giornale delle comunicazioni. Rivista mensile di Poste, Telegrafi e Telefoni, Roma, 1889-1891.

Secondary sources of the period

Adley, C. C. *Anglo-Indian Telegraphs or Communication with London, in Six Hours.* Calcutta: Sanders, Cones & Co, 1854.

Anderson, J. *Existing and Projected Telegraph Routes to India Considered.* London: Metchim & Son, 1868.

—. "Statistics of Telegraphy", *Journal of the Statistical Society of London* 35/3 (1872): 272-326.

Atlantic Telegraph Company. *A History of Preliminary Experimental Proceedings and a Descriptive Account of the Present State & Prospects of the Undertaking*. London: Jarrold and Sons, 1857.

Atlantic Telegraph: Its History, from the Commencement of the Undertaking in 1854, to the sailing of the ""Great Eastern" in 1866. London: Bacon and Co, 1866.

Bentham, J. "On the Liberty of the Press, and Public Discussion, Addressed to the Spanish People". *The works of Jeremy Bentham*, edited by J. Bowring. Edinburgh: William Tait, 1839.

Blavier, E. E. *Cours théorique et pratique de télégraphie électrique.* Paris : De Lacroix-Comon, 1857.

—. *Nouveau traité de télégraphie électrique.* Paris : Eugene Lacroix, 1866-67.

Blundell, J. W. *Manual of Submarine Telegraph Companies*, London: Published by the Author, 1872.

Breguet, L. *Manuel de télégraphie électrique*, Paris : L. Hachette,1862.

Bright, C. *Submarine Telegraphs. Their history, construction, and working*. London: Crosby Lockwood and Son, 1898.

—. *Imperial Telegraphic Communication.* London: P. S. King & Son, 1911.

Bois, V. *La télégraphie électrique*. Paris : Librarie L. Hachette, 1853.

Bond, R. *The Handbook of the Telegraph. Being a Manual of Telegraphy, Telegraph Clerks' Remembrances and Guide to Candidates for Employment in the Telegraph Service*. London: Virtue Brother and Co, 1864.

—. *Handbook of the Telegraph. Being a manual of telegraphy, telegraph clerks remembrances, and guide to candidates for employment in the telegraph service*. London: Lockwood & Co, 1873.

Bureau International des Administrations Télégraphiques. *La legislation télégraphique*. Berne: Imprimerie Rieder & Simmen, 1876.

Carême, F. and J. Raynaud, *Le télégraphe automatique de sir Charles Wheatstone*. Paris: Dunod, 1876.

Chadwick, E. "On the economy of telegraphy as part of a public system of postal communication". *Journal of the Society of Arts*, 1 March 1867.

—. "On the Railway Reform in Connection with a Cheap Telegraphic Post and a Parcel Post Delivery". *Journal of the Society of Arts*, 18 October 1867.

Clark, L. *Birmingham Wire Gauge. A Paper Read at the British Association for the Advancement of Science at Dundee, September 1867*. London E. & Spon, 1867.

Crotch, A. *The Hughes and Baudot Telegraphs*. London: S. Rentell & Co Limited, 1908.

Culley, R. S., *A Handbook of Practical Telegraphy*. London: Longman, Green, Roberts and Green, 1867.

—. *A Handbook of Practical Telegraphy*. London: Longmans, Green and Co, 1885.

Dood, G. *Railways, Steamers and Telegraphs. A glance at their recent progress and present state*. London-Edinburgh: W. & R. Chambers, 1868.

Douglas, J. C. *A Manual of Telegraph Construction: the Mechanical Elements of Electric Telegraph Engineering*. London: Charles Griffin and company, 1877.

Du Moncel, T. *Considérations Nouvelles sur l'électromagnétisme et ses applications*, Paris: J. Claye, 1853.

—. *Traité theorique et pratique de télégraphie eléctrique, a l'usage des employes télégraphistes, des ingenieurs, des constructeurs et des inventeurs*. Paris: Gauthier-Villars, 1864.

Fahie, J. J. *A History of Electric Telegraphy to the Year 1837*. London: E & F. N. Spom, 1884.

Field, H. M. *The Story of the Atlantic Telegraph*. New York: Charles Scribner's sons, 1893.

Figuier, L. *Exposition et Histoire des Principales Découvertes Scientifiques Modernes*. Paris : Langlois et Leclerq-Victor Masson, 1851.

Fleming, J. A., *Fifty Years of Electricity. The Memories of an Electrical Engineer*. London: Iliffe & Sons Limited, 1921.

Gavarret, J. *Télégraphie électrique*. Paris: Victor Masson et fils., 1861.

Gisborne, F. *Telegraphic Communication with India*. London: Edward Stanford, 1868.

Goldsmid, F. J., *State and Prospects of the Existing Indo-European Telegraph: Being a Reply to Recent Public Assertions on the Subject of Telegraphic Communication with India*. London, Edward Stanford, 1868.

—. *Telegraph and Travel. A Narrative of the Formation and Development of Telegraphic Communication between England and India, under the Orders of her Majesty's Government, with Incidental Notices of the Countries Traversed by the Lines*. London: McMillan and Co., 1874.

Hand-book to Electric Telegraph. London: W. Scales, Shoreditch, Clarck Warwick Lane, 1847.

Highton, E. *The Electric Telegraph: its History and Progress*, London: John Weale, 1852.

Hoskiaer, V. *Laying and Repairing of Electric Telegraph Cables*. London: E. &F. N. SPON, 1878.

Jevons, W. S. *Methods of Social Reforms and other papers*. London: MacMillan and Co., 1883.

Lardner, D. *The Electric Telegraph Popularised*. London: Walton and Maberly, 1855.

Laurencin, P. *La télégraphie terrestre, sous marin, pneumatique*. Paris : J. Rothschild, 1877.

Marqfoy, G. *Des reformes nécessaires en télégraphie*. Paris: Guillaumin, 1866.
Mercadier, E. *Traité élémentaire de télégraphie électrique*. Paris : G. Masson, 1880.
Miege, B. *Vade-mecum pratique de télégraphie électrique a l'usage des employés du télégraphe*. Paris : J. Hetzel et C.ie, 1855.
Mill, J. S. *Principles of Political Economy, Vol. II*. Boston: Charles C. Little & James Brown, 1848.
Newall, R. S. *Facts and Observations Relating to the Invention of the Submarine Cable and the Manufacture and Laying of the First Cable between Dover and Calais in 1851*. London: E. & F. Spon, 1882.
Ogilvie, A. M. "The State in Its Relations to Telegraphs and Telephones." *The Economic Journal* 18, no 72 (1908) : 602-609.
Phipson, T. *A Manual of the Various Electro-Magnetic Telegraphs at Present in Use*. London: J. Gilbert- J. M'Combe, 1847.
Preece, W. H. and J. Sivewright. *Telegraphy*. London: Longmans Green and Co, 1876.
Prescott, G. B. *History, Theory and Practice of the Electric Telegraph*. Boston: Ticknor and Fields, 1860.
Progress, P. *The Rail and the Electric Telegraph comprising a Brief History of Former Modes of Travelling and Telegraphic Communication*. London: R. Yorke Clarke and Co., 1848.
Sabine, R. *The Electric Telegraph*. London: Virtue Brothers & C., 1867.
—. *History and Progress of the Electric Telegraph*. London: Lockwood & C., 1872.
Saveney, E. "La télégraphie internationale", *Revue des deux mondes* (1872) : 360-384.
Saward, G. *The Trans-Atlantic Submarine Telegraph: a Brief Narrative of the Principal Incidents in the History of the Atlantic Telegraph Company*. London: printed for private circulation, 1878.
Schwendler, L. *Instructions for Testing Telegraph Lines and the Technical Arrangements of Offices*. London: Trubner & C., 1878.
Seeligmann, T. *Indian Rubber and Gutta percha: a Complete Practical Treatise on Indian Rubber and Gutta percha in their Historical, Botanical, Arboricural, Mechanical, Chemical and Electrical Aspects*. London: Scott Greenwood, 1903.
Shaffner, T. P. *The Telegraph Manual: a Complete History and Description of the Semaphoric, Electric and Magnetic Telegraphs of Europe, Asia, Africa and America, Ancient and Modern*. New York: Pudney & Russel, 1859.

Smith, W. *A Résumé of the Earlier Days of Electric Telegraphy*. London: Hayman Brothers and Lilly, 1881.
—. *The Rise and Extension of Submarine Telegraphy*: London: J. S. Virtue & Co, 1891.
Suarez Saavedra, A. *Tratado de telegrafia, Historia universal de la telegrafia*. Barcelona: Imprenta de Jaime Jepus, 1880-82.
Ternant, L. A. *Manuel pratique de télégraphie sous marine*. Paris : E. Lacroix, 1869.
—. *Transmission des signaux par les câbles sous-marins*. Parigi : Ducher, 1875
—. *Le siphon-recorder et le curb-sender automatique.* Paris : G. Masson, 1882.
Turnbull, L. *The Electro-magnetic Telegraph*. Philadelphia: A. Hart, 1853.
Walker, C. V. *Electric Telegraph Manipulation*. London: George Knight, 1850.
West, C. *The Story of my Life, by the Submarine Telegraph*. London: John Capman and company, 1859.
Wilson, G. *Electricity and the Electric Telegraph*. London: Longman, Brown, Green and Longmans, 1852.
—. *The Progress of the Telegraph*. Cambridge: MacMillan and Co., 1859.

Contemporary monographs and articles

Ahvenainen, J. *The Far Eastern Telegraphs: the History of Telegraphic Communications between the Far East, Europe and America before the First World War*. Helsinki: Finnish Academy of Science and Letters, 1981.
—. "Telegraphs, trade and policy. The role of the international telegraphs in the years 1870-1914." In *Commercial and Financial services*, edited by R. C. Michie, 505-515. Oxford and Cambridge, MA; Basil Blackwell, 1994.
—. *The European Cable Companies in South America before the First Words War*. Helsinki: Finnish Academy of Science and Letters, 2004.
—. "The International Telegraph Union: the Cable Companies and the Governments." In *Communications under the Seas: the Evolving Cable Network and its Implications* edited by B. Finn and D. Yang, 61-79. Cambridge, MA, London: The MIT Press, 2009.
Altermatt, C. "On Special Mission: Switzerland and its Diplomatic System." In *The Diplomats' World. A Cultural History of Diplomacy, 1815-1914*, edited by M. Mösslang and T. Riotte, 317-344. Oxford: Oxford University Press, 2008.

de Andrade Martins, R. "Resistance to the Discovery of electromagnetism: Oersted and the symmetry of the magnetic field." In *Volta and the History of Electricity*, edited by F. Bevilaqua and E. Giannetto, 245-265. Pavia/Milano: Università degli Studi di Pavia-Hoepli, 2003.

Appleyard R. *Pioneers of Electrical Communications*. London: MacMillan and Co, 1930.

Badenoch A., and A. Fickers. "Introduction: Europe Materializing? Toward a Transnational History of European Infrastructures." In *Materializing Europe: Transnational Infrastructures and the Project of Europe*, edited by A. Badenoch and A. Fickers, 1-23. Basingstoke, Hampshire and New York: Palgrave Macmillan, 2010.

Bayly, C. A. *The Birth of the Modern World 1780-1914. Global Connections and Comparisons*. Malden, Oxford, Victoria: Blackwell Publishing, 2004.

Balbi, G. "Studying the Social History of Telecommunications. Between Anglophone and Continental Tradition." *Media History* 15, no 1 (2009): 85-101.

—. *Le origini del telefono in Italia. Politica, economia, tecnologia e società*. Milan: Bruno Mondadori, 2011.

—. "The Origins of the Telephone in Italy, 1877–1915: Politics, Economics, Technology and Society." *International Journal of Communication* 5 (2011): 1058-1081.

—. "Telecommunications." In *Handbook of Communication History*, edited by P. Simonson, J. Peck, R. T. Craig, and J. P. Jackson, 209-222. London and New York: Routledge, 2012.

—. *Network Neutrality. Switzerland's Role in the Genesis of the Telegraph Union, 1855-1875*. Bern: Peter Lang, 2014.

Balbi, G., Calvo S., Fari, S., and G. Richeri. "'Bringing Together the Two Large Electric Currents that Divide Europe': Switzerland's Role in Promoting the Creation of a Common European Telegraph Space, 1849-1865." *ICON* 15: 61-80.

Balbi, G., Fari, S., Calvo S., and G. Richeri. "Swiss Specialties: Switzerland's Role in the Genesis of the Telegraph Union, 1855-1875." *Journal of European Integration History* 19, no 2 (2013): 207-225.

Balbi, G., Fari, S., Calvo S., and G. Richeri. "Specialità svizzere. L'influenza della Confederazione elvetica sull'origini dell'Unione Telegrafica, 1855-1875." *Tst: Transportes, Servicios y telecomunicaciones* 25 (2013): 150-175.

Balbi, G. and Kittler J. *Dialoguing With Socrates or Disseminating Like Jesus? Rereading Communication History Through 'One-to-One' and*

'One-to-Many' Lenses, Phoenix: ICA (International Communication Association), 2012.

Barabasi A. L. *Linked: The New Science of Networks*. Cambridge Massachusetts: Perseus Publishing, 2002.

Barbier, F. et C. Bertho-Lavenir. *Histoire des medias: de Diderot à Internet*. Paris: A. Colin, 2003.

Barton R. N. "New Media. The birth of telegraphic news in Britain 1847-68." *Media History* 16, no 4 (2010): 379-406.

Beauchamp, K. *History of Telegraphy*. IEE: Stevenage, 2001.

Bektas Y. "The Sultan's Messenger: Cultural Constructions of Ottoman Telegraphy, 1847-1880." *Technology and Culture* 41, no 3 (2000): 669-696.

Bell D. S. A. "Dissolving Distance: Technology, Space, and Empire in British Political Thought, 1770-1900." *The Journal of Modern History* 77 (2005): 523-562.

—. "Empire and international relations in Victorian political thought." *The Historical Journal* 49, no 1 (2006): 281-298.

Beniger J. *The Control Revolution. Technological and Economic origins of the Information Society*. Cambridge MA: Harvard University Press, 1989.

Bergier, J-F. *Wirtschaftgeschichte der Schweiz*. Zürich: Benziger Verlag, 1983.

Bertho-Lavenir, C. (ed.). *Histoire des télécommunications en France*. Toulouse: Eres, 1984.

Blackbourn, D. *The Long Nineteenth century: a History of Germany, 1780–1918*. New York: Oxford University Press, 1998.

Blatherwick, D. E. S. *The International Politics of Telecommunications*. Berkeley, CA: Institute of International Studies, University of California, 1987.

Boyce R. W. D. "Submarine Cables as a Factor in Britain's Ascendancy." In *Kommunikationsrevolutionen. Die neunen Medien des 16. und 19. Jahrhunderts* edited by M. North, 81-100. Köln-Weimar-Wein: Böhlau Verlag, 1995.

—. "Imperial Dramas and National Realities: Britain, Canada and the Struggle for a Pacific Telegraph Cable, 1879-1902." *English Historical Review* 115, no 460 (2000): 39-70.

Bregni, S. "A Historical Perspective on Telecommunications Network Synchronization." *IEEE Communications Magazine* 36, no 6 (1998): 158-166.

Brooks, J. *Telephone. The First Hundred Years*. New York & London: Harper & Row, 1972.

Brown, R. *Telecommunications. The Booming Technology*. New York: Doubleday & Company, Inc., 1970.
Bühlmann, E. *La ligne Siemens: la construction du télégraphe indo-européen 1867-1870*. Frankfurt : Peter Lang, 1999.
Buliung R. N. "Wired People in Wires Places: Stories about Machines and the Geography of Activity." *Annals of the Association of American Geographers* 101, no 6 (2011): 1365-1381.
Burns, R. *Communications: an International History of the Formative Years*. London: IEE, 2004.
Burtz L. and E. Hummel. "Standard Setting in International Telecommunications." *Telecommunications Policy* 8, no 1 (1984): pp. 3-6.
Calvo, S., Balbi, G., Fari S., and G. Richeri. "La voie suisse aux télécommunications. Politique, économie, technologie et société (1850–1915)." *Revue Suisse d'Histoire* 61, no 4 (2011): 435-453.
Carrè, P. A. "Archeologie d'une Europe des télécommunications." *Revue Française des Télécommunications* 70 (1989): 72-83.
Choudhury D. K. L. *Telegraphic Imperialism: Crisis and Panic in the Indian Empire, c. 1830-1920*. Basingstoke: Palgrave Macmillan, 2010.
Clark K., *International Communications. The American Attitudes*. New York: Columbia University Press, 1931
Coates V. T., Finn B. *A Retrospective Technology Assessment: Submarine Telegraphy. The transatlantic cables of 1866*. San Francisco: San Francisco Press, 1979.
Codding, G. A. Jr. *The International Telecommunication Union. An experiment in international cooperation*. Leiden: E. J. Brill, 1952.
—. *The Universal Postal Union: Coordinator of the International Mails*. New York: New York University Press, 1964.
Codding, G. A. and A. M. Rutkowski. *The International Telecommunication Union in a Changing World*. Deedham: Artech House, 1982.
Cookson, G. *The Cable: the Wire that Changed the World*. Stroud: Tempus, 2003.
—. "Submarine Cables: Novelty and Innovations 1850-1870." *Transactions of the Newcomen Society*, 76, no 2 (2006): 207-220.
Cornwall M. (ed.) *The Last Years of Austria-Hungary: a Multi-national Experiment in Early Twentieth-Century Europe*. Exeter: University of Exeter Press, 1990.
Cowhey, P. F. "The International Telecommunications Regime: The Political Roots of Regimes for High Technology." *International Organization* 44, no 2 (1990): 169-199.

David, P. A. "Clio and the Economics of QWERTY". *American Economic Review* 75, no 2 (1985): 332-337

—. "Path Dependence: a Foundational Concept for Historical Social Science." *Cliometrica: The Journal of Historical Economics and Econometric History* 1, no 2 (2007): 145-175.

Dent, N. J. H. "The Tensions in Liberalism." *The Philosophical Quarterly* 153, no 38 (1988): 481-485.

Descalzi, A. "Creación y desarrollo de la Unión Internacional de telecomunicaciones." In *Las comunicaciones entre Europa y América, 1500-1993*, edited by B. Magro, M. Llorente, O. Carvajal, 287-298. Palacio de Congresos de Madrid: Actas del I Congreso Internacional de Comunicaciones, 30 de noviembre – 3 de diciembre.

Devin, G. and M. C. Smouts. *Les organisations internationales*. Paris: Colin, 2011.

Dibner, B. *The Atlantic Cable*. Norwalk. CT: Burndy Library, 1959.

—. *Oersted and the Discovery of Electromagnetism*. New York: Blaisdell, 1962.

Direction Générale des PTT (publ. par). (publ. par), *Un siècle de télécommunications en Suisse, 1852-1952. Tome I, Télégraphe*. Berne, 1952.

—. *Un siècle de télécommunications en Suisse, 1852-1952. Tome II, Téléphone, sources de courant et installations d'énergie, lignes aériennes et lignes souterraines*. Berne, 1959.

—. *Un siècle de télécommunications en Suisse, 1852-1952. Tome III, Téléphone (suite et fin)*. Berne, 1963.

Doyle, M. W. "Liberalism and World Politics." *The American Political Science Review* 4, no 80 (1986): 1151-1169.

Dordick, H. S. "The Origins of the Universal Services: History as a Determinant of Telecommunications Policy." *Telecommunications Policy* 14 (1990): 223-231.

Du Boff, R. B. "The Telegraph in Nineteenth-Century America: Technology and Monopoly." *Comparative Studies in Society and History* 4, no 26 (1984): 571-586.

Dunsheath P. *A History of Electrical Engineering*. London: Faber and Faber, 1962.

Durand Barthez, P. *Union Internationale des Télécommunications*. Thèse pour le doctorat en droit, Université de Paris I-Pantheon-Sorbonne Sciences Economiques-Science Humaines-Sciences Juridiques, 1979.

Dzur, A. W. "Nationalism, Liberalism and Democracy". *Political Research Quarterly*, 22, no 1 (2002): 191-211.

Ehs, T. "An Unwritten History: The Europeanisation of Switzerland." Paper presented at the EUSA 10th Biennial Conference, Montreal, Canada, May 17-19, 2007.

—. "History in Europeanisation Studies: Lessons from Switzerland." *Contemporary European Studies* 3, no 1 (2008): 23–39.

Fari, S. "Technology on the Wire. Technological Changes in the First Thirty Years of the Italian Telegraph Experience: Achievements and Difficulties." In *Communication and its Lines. Telegraphy in the 19th Century among Economy, Politics and Technology*, edited by A. Giuntini, 135-158. Prato: Istituto di Studi Storici Postali, 2004.

—. "Uccide più la parola che la spada. Telecomunicazioni e questioni militari nell'Italia del XIX secolo." *Ricerche Storiche* 1 (2006): 5-28.

—. *Una penisola in comunicazione. Il servizio telegrafico italiano dall'Unità alla Grande Guerra*. Bari: Cacucci Editore, 2008.

Fari S., Balbi G. and G. Richeri. "A Common Technical Culture of Telegraphy: the Telegraph Union and the Significance of Technological Standardization, 1865-1875." Paper presented at the HISTELCON Conference, Pavia, September 5-7, 2012 (available http://ieeexplore.ieee.org/xpl/articleDetails.jsp?tp=&arnumber=6487580&queryText%3DHistelcon).

Fari, S., Balbi, G., and G. Richeri. "Telecommunications Italian Style: The Shaping of the Constitutive Choices (1850-1914)". In *Italian Technology from the Renaissance to the Twentieth Century* edited by A. Guagnini and L. Molà, 235-258. "History of Technology" (book-series) n. 32. London-New Delhi-New York-Delhi: Bloomsbury, 2014.

Fari, S., Balbi, G., and G. Richeri. "The Bureaucratisation of the Telegraph Union". *Storia Economica*, XVI, no 2 (2013): pp. 377-394.

Fari S. "Financing Telegraph Infrastructure: the Case of Great Britain (1850-1900)." In *Infrastructure Finance in Europe: Insights into the History of Water, Transport and Telecommunication* edited by Y. Cassis, G. De Luca and M. Florio. Oxford: Oxford University Press, forthcoming.

Feldman, M. L. B. *The United States in the International Telecommunication Union and in Pre-ITU conferences: Submarine Cables, Overland Telegraph, Sea and Land Radio, Telecommunications*. Baton Rouge: printed autonomously, 1974.

Ferderer, J. P. "Advances in Communication Technology and Growth of the American Over-the-Counter Markets, 1876-1929." *The Journal of Economic History* 68, no 2 (2008): 501-534.

Fickers, A., and P. Griset. *Eventing Europe: Electronic Information and Communication Spaces in Europe, 1850-2000.* London: Palgrave/MacMillan, forthcoming.

Fickers A., and S. Lommers. "Eventing Europe: Broadcasting and the Mediated Performances of Europe." In *Materializing Europe: Transnational Infrastructures and the Project of Europe*, edited by A. Badenoch and A. Fickers, 225-251. Basingstoke, Hampshire and New York: Palgrave Macmillan, 2010.

Field, A. J. "The Magnetic Telegraph, Price and Quantity Data, and the New Management of Capital." *The Journal of Economic History*, 52, no 2 (1992): 401-413.

Finn, B. "Submarine Telegraphy: a Study in Technical Stagnation." In *Communications under the Seas: the Evolving Cable Network and its Implications*, edited by B. Finn and D. Yang, 9-24. Cambridge, MA and London: The MIT Press, 2009.

Flichy, P. *Une histoire de la communication moderne: espace public et vie privée*. Paris: La Decouverte, 1991.

Foreman-Peck, J. "L'état et le développement du réseau de télécommunications en Europe à ses débuts." *Histoire, économie et société* 3 (1989): 383-402.

Franksen, O. I. *H. C. Oersted – a Man of the two Cultures*. Birlerod: Strandsberg Forlag, 1981.

Friedrich A. "Metaphorical Anastomoses: The Concept of "Network" and its Origins in the Nineteenth Century." *Travelling Concepts for the Study of Culture* edited by B. Neumann and A. Nünning, 119-144. Berlin/Boston: Walter de Gruyter, 2012.

Garbade, K. and W. L. Silber. "Technology, Communication and the Performance of Financial markets: 1840-1975." *Journal of Finance* 33, no 3 (1978): 819-832.

Gassler C. "Some Problems in the Development of the Communications Industry". *The American Economic Review* 35, no 4 (1945): 585-606.

Genschel, P. and R. Werle. "From National Hierarchies to International Standardization: Modal Changes in the Governance of Telecommunications." *Journal of Public Policy* 13, no 3 (1993): 203-225.

Gitelman L. *Always Already New: Media, History and the Data of Culture*. Cambridge MA: MIT Press, 2006.

Gitelman L. and G. B. Pingree (eds.). *New Media, 1740-1915*. Cambridge MA: MIT Press, 2003.

Giuntini, A. 2011. *Le meraviglie del mondo: il sistema internazionale delle comunicazioni nell'Ottocento*. Prato: Istituto di studi storici postali, 2011.

Gneme, G. *Della «Unione Internazionale delle Telecomunicazioni» e della «Unione Postale Universale»*. Roma: Poligrafico dello Stato, 1941.

Greig J. M. "The End of Geography? Globalization, Communications, and Culture in the International System." *Journal of Conflict Resolution* 46, no 2 (2002): 225-243.

Griset P. *Entreprise, technologie et souveraineté : les télécommunications transatlantiques de la France*. Paris : Editions Rive Droite, 1996.

Grossi, V. "Le rôle international de personnalités suisses du XIXe siècle dans le domaine des télégraphes." *Hispo,* Octobre 1984: 43-50.

—. "Technologie et diplomatie suisse au XIXe Siècle." *Relations internationales* 39 (1984): 287-307.

Habermas, J. *The Structural Transformation of the Public Sphere: an Inquiry into a Category of Bourgeois Society*. Cambridge MA: MIT Press, 1991.

Hamelink, C. J. *The Politics of World Communication: a Human Rights Perspective*. London and Thousand Oaks, CA: Sage, 1994.

Hamill L. "The Social Shaping of British Communications Networks prior to the First World War." *Historical Social Research* 35, no 1 (2010): pp. 260-286.

Hampf, M., and S. Müller-Pohl (eds.). *Global Communication Electric. Business, News and Politics in the World of Telegraphy*. Frankfurt and New York: Campus, 2013.

Hannah, L. "Logistics, Market Size, and Giant Plants in the Early Twentieth Century: a Global View." *The Journal of Economic History* 68, no 1 (2008): 46-79.

Harper J. *Monopoly and Competition in British Telecommunications. The Past, the Present and the Future*. London: Pinter, 1997.

Harris C. P. "The Persian Gulf Submarine Telegraph of 1864." *The Geographical Journal* 135, no 2 (1969): 169-190.

Harris, D. "European Liberalism in the Nineteenth Century." *The American Historical Review* 60, no 3 (1955): 501-526.

Hazlewood, A. "The Origin of the State Telephone Service in Britain." *Oxford Economic Papers, new series* 5, no 1 (1953): 13-25.

Headrick, D. R. *The Tools of Empire: Technology and European Imperialism in the Nineteenth Century*. New York: Oxford University Press, 1981.

—. *The Tentacles of Progress: Technology Transfer in the Age of Imperialism, 1850-1940*. New York and Oxford: Oxford University Press, 1988.

—. *The Invisible Weapon. Telecommunications and International Politics 1851-1945* New York: Oxford University Press, 1991.

—. *When Information Came of Age. Technologies of Knowledge in the Age of Reason and Revolution, 1700-1850*. London & New York: Oxford University Press, 2000.

Headrick, D. R. and P. Griset. "Submarine Telegraph Cables: Business and Politics, 1838-1939." *Business History Review* 75, no 3 (2001): 543-578.

Heilbroner, R. "Do Machines Make History?" *Technology and Culture* 8, no 3: 335-345.

Helgesson, C.-F. *Making a Natural Monopoly: the Configuration of a Techno-Economic Order in Swedish Telecommunications*. Stockholm: Stockholm School of Economics, the Economic Research Institute, 1999.

Hempstead C. A. "The Early Years of Oceanic Telegraphy: Technology, Science and Politics." *IEE Proceedings* 136, no A6 (1989): 297-305.

—. "Representations of Transatlantic Telegraphy." *Engineering Science and Education Journal* (1995): pp. S17-S25.

Henry K. *Liberalism and the Culture of Security*. Tuscaloosa: The University of Alabama Press, 2011.

Herren, M. *Internationale Organisationen seit 1865. Eine Globalgeschichte der internationalen Ordnung*. Darmstadt: Wissenschaftliche Buchgesellschaft, 2009.

—. "Governmental Internationalism and the Beginning of a New World Order in the Late XIX Century." In *The Mechanics of Internationalism: Culture, Society, and Politics from the 1840s to the First World War*, edited by M. H. Geyer and J. Paulmann, 121-144. Oxford: Oxford University Press, 2001.

Hershey, A. S. "History of International Law Since the Peace of Westphalia." *The American Journal of International Law* 6, no 1 (1912): 30-69.

Hills, J. *The Struggle for Control of Global communication: the Formative Century*. Urbana: University of Illinois Press, 2002.

Hoag, C. "The Atlantic Telegraph Cable and Capital Market Information Flows." *Journal of Economic History* 66, no 2 (2006): 342-53.

Hobsbawm, E. *Age of Capitalism 1848-1875*. London: Hachette, 2010.

Hochfelder, D. "A Comparison of the Postal Telegraph Movement in Great Britain and the United States, 1866-1900." *Enterprise & Society* 1 (2000): 739-761.

Hofmann, M. "The Telephone and Its Significance to Switzerland's First Subscribers." *Bulletin Technique PTT* 10 (1980): 390-392.

Horrenberger, J. *L'Union Internationale des Télécommunications ou Les exigences techniques comme factor de la coopération internationale.* Mémoire pour l'obtention du diplôme des Hautes Etudes Européens section des sciences de l'information, Université de Strasbourg, 1976.

Hugill, P. J. *Global Communications since 1844: Geopolitics and Technology*. Baltimore: Johns Hopkins University Press, 1999.

Huurdeman, A. A. *The Worldwide History of Telecommunications*. London: IEE, 2003.

Imlah, A. G. *Britain and Switzerland 1845-60: a Study of Anglo-Swiss Relations during some Critical Years for Swiss Neutrality*. Hamden (Connecticut): Archon Books, 1966.

Iriye, A. P. and Y. Saunier (eds.). *The Palgrave Dictionary of Transnational History*. Basingstoke UK and New York: Palgrave Macmillan, 2009.

ITU. *L'Union télégraphique international: 1865-1915*. Berne: Bureau international de l'Union télégraphique, 1915.

—. *L'Union Internationale des télécommunications, cent ans de coopération international*. Genève, 1965.

John, R. R. "Private enterprise, public good? Communications deregulation as a national political issue, 1839-1851." In *Communication and its Lines. Telegraphy in the 19th Century Among Economy, Politics and Technology*, edited by A. Giuntini, 33-56. Prato: Istituto di Studi Storici Postali, 2004.

—. *Study on the Universal Postal Service and Postal Monopoly. Appendix D. History of Universal Service and the Postal Monopoly*. George Mason University: School of Public Policy, 2008.

—. "Telecommunications." *Enterprise and Society* 9 (2008): 507-520.

—. *Network Nation. Inventing American Telecommunications*. Cambridge MA and London: The Belknap Press of Harvard University Press, 2010.

Johnson P. *Making the Market. Victorian Origins of Corporate Capitalism*. Cambridge: Cambridge University Press, 2010.

Jost, H. U. "Origines, interprétations et usages de la 'neutralité helvétique'." *Matériaux pour l'histoire de notre temps* 93 (2009): 5-12.

Kahan, A. S. *Liberalism in Nineteenth-Century Europe : the Political Culture of Limited Suffrage*. Basingstoke and New York: Palgrave-Macmillan, 2003.

Kaukianen, Y. "Shrinking the World: Improvements in the Speed of Information Transmission, c. 1820-1870." *European Review of Economic History* 5, no 1 (2001), 1-28.

Kennedy, P. M. "Imperial Cable Communications and Strategy, 1870-1914." *The English Historical Review* 86, no 341 (1971): 728-752.

Kent D. "Containing Disorder in the Age of Equipoise: Troops, Trains and the Telegraph." *Social History* 38, no 3 (2013): 308-327.

Kieve, J. *The Electric Telegraph. A Social and Economic History*. Devon: David and Charles Newton Abbot, 1973.

Kinnebrock, S., C. Schwarzenegger and E. McLuskie (eds.). *What is Communication History? European Answers II*. Special Issue of *Medien & zeit* 3-4 (2011).

Kott, S. (ed.). *Une autre approche de la globalisation: Socio-histoire des organisations internationales (1900-1940*). Numéro spécial de *Critique internationale* 52 (2011).

Krasne, S. D. "Communications and National Power: Life on the Pareto Frontier." *World Politics* 43, no 3 (1991): 336-366.

Kriesi, H. *Le système politique suisse*. Paris: Economica, 1998.

Kusserow K. "Technology and Ideology in Daniel Huntigton's *Atlantic Cable Projectors*." *American Art* 24, no 1 (2010): 94-113.

L'Union internationale des télécommunications, 1865-1995. London: International Systems and Communications LTD, 1995.

Laborie, L. *La France, l'Europe et l'ordre International des communications, 1865-1959*. Thèse pour le Doctorat en histoire contemporaine, Université Paris IV-Sorbonne, 2006.

—. "En chair et en normes. Les participants aux conférences de l'Union internationale des télécommunications, de sa fondation à sa refondation (1865-1947)." *Flux. Cahiers scientifiques internationaux Réseaux et Territoires* 74 (2008): 92-98.

—. *L'Europe mise en réseaux. La France et la coopération internationale dans les postes et les télécommunications, années 1850-années 1950*. Bruxelles: PIE Peter Lang, 2010.

—. "De quoi l'universel est-il fait? L'Europe, les empires et les premières organisations internationales." *Les Cahiers Irice*, 9 (2012): 11-22.

Larmore, C. "Political Liberalism." *Political Theory* 18, no 3 (1990): 339-360.

Lequeux, J. *François Arago, un savant généreux*. Paris : EDP- Sciences, 2008.

Levinson P. *Soft Edge. A Natural History & Future of the Information Revolution*. London: Routledge, 1997.

Lew, B. and B. Cater. "The Telegraph, Co-ordination of Tramp Shipping, and Growth in World Trade, 1870-1910." *European Review of Economic History* 10, no 2 (2006): 147-73.

Lindley, L. G. "Watered Stock and Control of Telegraph Rates: Early Proposals for Regulating a Public Utility." *The Journal of Economic History* 32, no 1 (1972): 403-408.

Liniger-Goumaz, M. *La Suisse, sa neutralité et l'Europe*. Genève: Éditions du Temps, 1964.

Lyall, F. *International communications: the International Telecommunication Union and the Universal Postal Union*. Farnham, Surrey UK and Burlington, VT: Ashgate, 2011.

Lubrano, A. *The Telegraph: How Technology Innovation Caused Social Change*. New York: Garland, 1997.

MacPherson, A. *International Telecommunication Standards Organizations*. Norwood: Artech House, 1990.

Mance O. and J. E. Wheeler. *International Telecommunications*. Oxford: Oxford University Press, 1943.

Martin, L. "Interests, Power and Multilateralism." *International Organization* 46, no 4: 765-792.

Mc Closkey, H. T. "Mill's Liberalism." *The Philosophical Quarterly* 51, no 13 (1963): 143-156.

Mc Cormick, P. K. "Private sector influence in the International Telecommunication Union." *The Journal of Policy, Regulation and Strategy for Telecommunications, Information and Media* 9, no 4 (2007): 70-80.

McMahon, P. "Early Electrical Communications Technology and Structural Change in the International Political Economy. The Case of Telegraphy and Radio." *Prometheus: Critical Studies in Innovation* 20, no 4: 379-390.

Meyer, V. *L'Union Internationale des Télécommunications et son Bureau*. Bern: typewritten, ITU's Library, 1946.

Michaelis, A. R. *Du sémaphore au satellite*. Genève: ITU, 1965.

Michie R. C. "The London Stock Exchange and the British Securities Market, 1850-1914." *The Economic History Review* New Series 38, no 1 (1985) : 61-82.

Milivojević, M., and P. Maurer (eds.). *Swiss neutrality and security: armed forces, national defence, and foreign policy*. New York: Berg and St. Martin's Press, 1990.

Misa, T. J., and J. Schot. "Inventing Europe: Technology and the Hidden Integration of Europe. Introduction to the special issue." *History and Technology* 21, no 1 (2005): 1-20.

Mokyr, J. "Technological Inertia in Economic History." *The Journal of Economic History* 52, no 2 (1992): 325-338.

Montagné, J.C. *Histoire des moyens de Télécommunication de l'Antiquité à la Seconde Guerre Mondial: du sifflet préhistorique à la télévision. Les hommes, les idées*. Bagneux : J. C. Montagné, 1995.

Morrow, J. D. "Modelling the Forms of International Cooperation: Distribution Versus Information." *International Organization* 48, no 3 (1994): 387-423.

Morus, I. R. "The Electric Ariel: Telegraphy and Commercial Culture in Early Victorian England." *Victorian Studies* 39, no 3 (1996): 339-378.

—. "The Nervous System of Britain: Space, Time and the Electric Telegraph in the Victorian Age." *The British Journal for the History of Science* 33, no 4 (2000): 455-475.

Mueller, M. L. *Universal Service. Competition, Interconnection and Monopoly in the Making of the American Telephone System.* Cambridge MA: MIT/AEI, 1997.

Müller-Pohl, S. "By Atlantic Telegraph. A study on Weltcommunication in the 19th Century." *m&z* 4 (2010): 40-54.

—. "The Transatlantic Telegraphs and the Class of 1866 - the Formative Years of Transnational Networks in Telegraphic Space, 1858-1884-89." *Historical Social Research* 35, no 1 (2010): 237-259.

—. "Working the Nation State: Submarine Cable Actors, Cable Transnationalism and the Governance of the Global Media System, 1858-1914." In *The Nation State and Beyond. Governing Globalization Processes in the Nineteenth and Twentieth Century*, edited by I. Löhr and R. Wenzlhuemer. Berlin: Springer Verlag Berlin, 2013.

Müller-Pohl, S., and M. Hampf (eds.). *Global Communication Electric: Actors of a Globalizing World.* Frankfurt: Campus, 2013.

Nachszunow, G. *Développement des télécommunications et organisations internationales*. Créteil: W. Nachszunow, 1989.

Navarro, J. F. *ITU: síntesis de organización, funcionamiento y objetivos de la Union Internacional de Telecomunicaciones.* Chile: Entel-Chile, 1975.

Nickles, D. P. *Under the Wire: How the Telegraph Changed Diplomacy*, Cambridge, MA and London: Harvard University Press, 2003.

Noam, E. *Telecommunications in Europe*. New York: Oxford University Press, 1992.

Nonnenmacher, T. "Law, Emerging Technology, and Market Structure: the Development of the Telegraph Industry, 1838-1868." *The Journal of Economic History* 57, no 2 (1997): 488-490.

—. "State Promotion and Regulation of the Telegraph Industry, 1845-1860." *The Journal of Economic History*, 61, no 1 (2001): 19-36.

—. "Network Quality in the Early Telegraph Industry." *Research in Economic History* 23 (2005): 61-82.

Nye, D. E. "Shaping Communication Networks: Telegraph Telephone Computer." *Social Research* 64, no 3 (1997): 1067-1091.

Olivé Roig, S. *Historia de la telegrafía óptica en España*. Madrid: Ministerio de Trasporte, Turismo y Comunicaciones, 1990.

—. *El nacimiento de la telecomunicación en España*. Madrid: Cuadernos de Historia de telecomunicaciones nº 4, Escuela Técnica Superior de Ingenieros de Telecomunicación, 2004.

Overman M. *Understanding Telecommunications*. Portsmouth: Eyre & Spottiswide, 1974.

Park, D. W. Janowsky N. W and Jones S. (eds.). *The Long History of New Media: Technology, Historiography and Newness in Context*. New York: Peter Lang, 2011.

Perrin, G. *La neutralité permanente de la Suisse et les organisations internationales*. Heule: UGA, 1964.

Perry, C. R. *The Victorian Post Office. The growth of a bureaucracy*. Woodbridge: The Royal Historical Society/The Boydell Press, 1992.

—. "The Rise and Fall of Government Telegraphy in Britain." *Business and Economic History* 26, no 2 (1997): 416-425.

Peters A., and S. Peter. "International Organizations: Between Technocracy and Democracy." In *The Oxford Handbook of the History of International Law*, edited by B. Fassbender and A. Peters, 170-197. Oxford: Oxford University Press, 2012.

Pflieger, G. "La Suisse est-elle un espace première classe?" *Flux* 72, no 2-3 (2008): 4-12.

Pinch, T. and W. E. Bijker. "The Social Construction of facts and artefacts: or how the sociology of Technology might benefit each other." *Social Studies of Science* 14, no 3 (1984): 419-424.

Pitt, D. C. *The Telecommunications function in the British Post Office. A case study of Bureaucratic Adaptation*. Westmead, Farnborough: Saxon House, 1980.

Pomper P. "The History and Theory of Empires." *History and Theory* 44, no 4 (2005): 1-27.

Potter, S. J. "Webs, Networks, and Systems: Globalization and the Mass Media in the Nineteenth and Twentieth Century British Empire." *Journal of British Studies* 46, no 3 (2007): 621-646.

Pravato, S. "De Télécom PTT à Swisscom." In *Du monopole à la concurrence. Analyse critique de l'évolution de 6 entreprises suisses*, edited by M. Finger, S. Pravato and J. N. Rey. Lausanne: LEP, 1994.

Reindl, J. *Der Deutsch-Österreichische Telegraphenverein und die Entwicklung des Deutschen Telegraphenwesens 1850–1871*. Frankfurt am Main: Peter Lang, 1993.

—. "Partikularstaatliche Politik und Technische Dynamik: Die Drahtgebundene Telegraphie und der Deutsch-Österreichische Telegraphenverein von 1850." In *Vom Flügeltelegraphen zum Internet: Geschichte der modernen Telekommunikation*, edited by H.-J. Teuteberg and C. Neutsch, 27–46. Stuttgart: F. Steiner, 1998.

Renaud, J. L. *The Changing of the International Telecommunication Union : an Historical Analysis of the Development Assistance*. Dissertation submitted to Michigan State University in partial fulfilment of the requirements for the degree of Doctor of Philosophy, Mass Media PhD Program, 1986.

Richeri, G. "The Media Amidst the Enterprises, the Public and the State." *Studies in Communication Sciences* 6, no 2 (2006): 131-143.

Roach J. "Liberalism and the Victorian Intelligentsia." *Cambridge Historical Journal* 13, no 1: 58-81.

Roberts S. *Distant Writing. A History of the Telegraph Companies in Britain between 1838 and 1868*. London: Steven Roberts, 2006-2012. Unpublished in hardcopy but available on the Internet site: http: //distantwriting.co.uk/, accessed 8 August 2013.

Rosenberg, N. *Exploring the Black Box: Technology, Economics, and History*. Cambridge UK and New York: Cambridge University Press, 1994.

Ruggie, J. G. "Multilateralism: the Anatomy of an Institution." *International Organization* 46, no 3 (1992): 561-598.

Ruggie, J. G. (ed.) *Multilateralism Matters: Theory and Praxis of an Institutional Form*. New York: Columbia University Press, 1993.

Rutkowski, A., and W. P. Dizard. "The International Telecommunication Union and the United States: Partners or Rivals?" In *International Telecommunications and Information Policy: Papers from a Symposium (P.1) and a Report of the National Telecommunications Information Administration (NTIA) 1983 (P.2)*, 27-45. Washington: Communications Pr., 1984.

Sachs, J. D., A. Warner, A. Aslund and S. Fischer. "Reform and the Process of Global Integration." *Brookings Papers on Economic Activity* 1, no 1 (1995): 1-118.

Schipper, F., and J. Schot. "Infrastructural Europeanism, or the Project of Building Europe on Infrastructures." *History and Technology* 27. no 3 (2011): 245-264.

Scholl, L. U. "The Global Communication Industry and Its Impact on International Shipping before 1914." In *Global Markets: The Internationalization of the Sea Transport Industries since 1850*, edited by D. J. Starkey and G. Harlaftis, 195-215. St John's: International Maritime Economic History Association, 1998.

Schmidt, S. and R. Werle. *Coordinating Technology: Studies in the International Standardization of Telecommunications*. Cambridge-London: MIT Press, 1998.

Schneider, V. *Die Transformation der Telekommunikation. Vom Staatsmonopol zum globalen Markt, 1800-2000*. Frankfurt am Main: Campus, 2001.

Shahvar S. "Iron Poles, Wooden Poles: The Electric Telegraph and the Ottoman: Iranian Boundary Conflict, 1863-65." *British Journal of Middle Eastern Studies* 34, no 1 (2007): 23-42.

Shapiro S. J. "John Stuart Mill, Pioneer of Democratic Liberalism in England." *Journal of the History of Ideas* 2, no 4 (1944): 127-160.

Shiers, G. (ed.) *The Electric Telegraph. An Historical Anthology*. New York: Arno Press, 1977.

Showalter, D. "Soldiers into Postmasters? The Electric Telegraph as an Instrument of Command in the Prussian Army." *Military Affairs* 37, no 2 (1973): 48-52.

Silverstone, R. "What's New about New Media?" *New Media and Society* 1 (1999): 10-12.

Solymar, L. *Getting the Message. A History of Telecommunications*. New York: Oxford University Press, 1999.

Starr, P. *The Creation of the Media: Political Origins of Modern Communications*. New York: Basic Books, 2004.

Steel, G. J. *A Thread Across the Ocean. The Heroic Story of the Transatlantic Cable*. London: Simon & Schuster, 2002.

Sterling, C., Bernt, P., and M. B. H. Weiss. *Shaping American Telecommunications: a History of Technology, Policy and economics*. Mahwah, New York: Lawrence Erlbaum Associates, 2006.

Stoddart Coggeshall I. *An Annotated History of Submarine Cables and Overseas Radiotelegraphs: 1851-1934* (ed.: Donard de Cogan). Norwich: School of Information Systems (UEA), 1993.

Tarjanne, P. (avant-propos par). *Union internationale des télécommunications: célèbre 130 années d'existence 1865 – 1995*. Genève: International Telecommunication Union and International Systems and Communications Ltd., 1996.

Tegge, A. *Die Internationale Telekommunikations-Union: Organisation und Funktion einer Weltorganisation im Wandel.* Baden-Baden: Nomos Verlagsgesellschaft, 1994.

Thompson, J. B. *The Media and Modernity: a Social Theory of the Media.* Stanford, CA: Stanford University Press, 1995.

Trainer M. "The Role of Patents in Establishing Global Telecommunications." *World Patent Information* 29 (2007): pp. 352-362.

Ulrico José, H. "Origines, interprétations et usages de la 'neutralité helvétique." *Matériaux pour l'histoire de notre temps* 93 (2009): 5-12.

Van der Herten, B. and P. Verhoest. "La contribution belge à la création des réseaux internationaux de communication du XIX siècle." In *Les réseaux européens transnationaux XIXe-XXe siècle : quels enjeux ?*, edited by Michelle Merger, Albert Carreras and Andrea Giuntini, 33-47. Nantes : Ouest Editions, 1995.

Van der Vleuten, E. "Towards a Transnational History of Technology: Meanings, Promises, Pitfalls." *Technology and Culture* 49, no 4 (2008): 974-994.

Van der Vleuten, E., and A. Kaiser (eds.). *Networking Europe. Transnational Infrastructures and the Shaping of Europe, 1850–2000.* Sagamore Beach, MA: Science History Publications, 2006.

Wallsten S. "Returning to Victorian Competition, Ownership, and Regulation: An Empirical Study of European Telecommunications at the Turn of the Twentieth Century." *The Journal of Economic History* 65, no 3 (2005): 693-722.

Wenzlhuemer, R. "Editorial – Telecommunication and Globalization in the Nineteenth Century." *Historical Social Research / Historische Sozialforschung* 35, no 1 (2010): 7-18.

—. "Globalization, Communication and the Concept of Space in Global History." *Historical Social Research / Historische Sozialforschung* 35, no 1 (2010): 19-47.

—. "The History of Standardization in Europe". *European History Online (EGO)* (2010): URL: http://www.ieg-ego-eu/wenzlhuemer-2010-en, [04/05/2014]

—. *Connecting the Nineteenth-Century World. The Telegraph and Globalization*. Cambridge UK: Cambridge University Press, 2013.

Williams, M. B. *History and Evolution of Telecommunications*. London: typewritten, BT Library, 1974.

Wilson, G., *The Old Telegraphs*. London: Phillimore, 1976.

Winseck, D. R. "Globalizing telecommunications and Media history: beyond methodological nationalism and the struggle for control model of communication history." In *Global Communication Electric. Business, News and Politics in the world of Telegraphy* edited by M. Michaela Hampf and Simone Müller-Pohl, 35-62. Frankfurt/New York: Campus Verlag, 2013,

Winsek, D. R., and R. M. Pike. *Communication and Empire: Media, Markets, and Globalization, 1860-1930*. Durham: Duke University Press, 2007.

—. "Communication and Empire: Media Markets, Power and Globalization, 1860-1910." *Global Media and Communication* 4, no 1: 7-37.

Yates, J. "The Telegraph's Effect on Nineteenth Century Markets and Firms." *Business and Economic History* 15 (1986): 149-63.

Ziegler, D. "Particularisme et développement des réseaux de transport dans les États allemands (1815-1866)." In *Les réseaux européens transnationaux XIXe-XXe siècles quels enjeux ?* edited by Michèle Merger, Albert Carreras and Andrea Giuntini, 23-32. Nantes : Ouest Éditions, 1995.